IMAGES
of America

KEMMERER

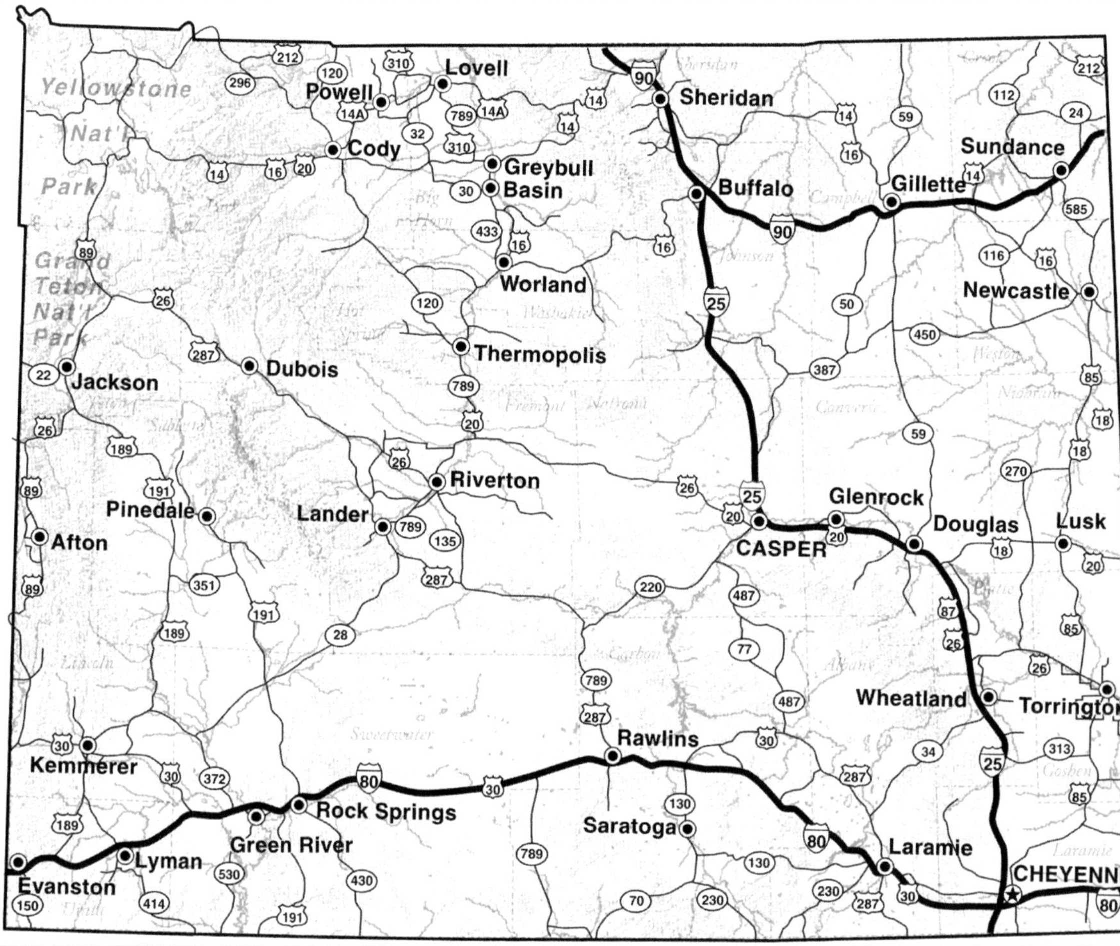

This map shows the location of Kemmerer in Lincoln County, located in southwest Wyoming. Kemmerer was initially part of Uinta County, one of Wyoming's original five counties, but in 1911, Uinta County was divided, and Kemmerer became the new county seat. The southwest corner of Wyoming is abutted by Idaho and Utah. Lincoln County is the only county in Wyoming named after a U.S. president. (Courtesy Fossil Country Museum.)

On the Cover: All early celebrations in Kemmerer were held in the Triangle Park in the center of town. The cover photograph shows the town children celebrating the Fourth of July in 1914. The children would march around the Triangle Park, and each child would receive a bag containing candy, nuts, and an orange. Horse racing, baseball games, footraces, and fireworks were enjoyed at this annual celebration. (Courtesy Fossil Country Museum.)

IMAGES of America

KEMMERER

Judy Julian and the Fossil Country Museum

ARCADIA PUBLISHING

ISBN 978-0-7385-6900-0

Published by Arcadia Publishing
Charleston, South Carolina

Printed in the United States of America

Library of Congress Control Number: 2008941501

For all general information, please contact Arcadia Publishing:
Telephone 843-853-2070
Fax 843-853-0044
E-mail sales@arcadiapublishing.com
For customer service and orders:
Toll-Free 1-888-313-2665

Visit us on the Internet at www.arcadiapublishing.com

This book is dedicated to the coal miners, power plant and gas plant employees, and ranchers who have kept Kemmerer alive for over 100 years.

Contents

ACKNOWLEDGMENTS

I am fortunate to have the support and encouragement of my family in writing this book. My husband, Mike, and my parents, Jim and Linda Roberts, were able to help fill in some of the missing pieces of the puzzle from family or firsthand knowledge. The Fossil Country Museum archives were the largest source of information. Nancy Peternal, founder of the Fossil Country Museum, had the foresight to find a building to house all this information and these photographs. I would like to thank my mother and my daughter, Kassie, for helping me dig through all the information and Kay Hinton for proofreading and correcting my mistakes. I would also like to thank the Fossil Country Museum for providing me with the privilege of writing this book. All the photographs shown in this book belong to the Fossil Country Museum.

Introduction

Kemmerer, like most small Western towns, has a rich and significant history, starting with fur trapping, ranching, railroads, and coal mining. It is significant for being the home of the first J. C. Penney store, the gateway to Fossil Butte National Monument, and the headquarters for the nation's deepest open-pit coal mine.

Kemmerer has been a coal mining community for over 100 years, and as with many coal mining towns, it was typical of the development of the coal mining industry in the West. Mining has played a major role in the past and present success of Kemmerer. The permanence of the area's mineral resources is evident in the quality and nature of early buildings; most were made of native stone quarried from the local area. This signaled a permanent atmosphere to the town.

With the arrival of the transcontinental railroad to Wyoming in 1869, the need for coal arose to run the engines. In July 1897, the Oregon Short Line Railroad was built through the Kemmerer area. This was a spur of the Union Pacific Railroad that established a direct route to the Pacific Northwest. The railroad was the main reason for the existence of coal towns.

The town of Kemmerer was established in 1897 by Patrick Quealy and Mahlon Kemmerer, for whom the town is named. The town site was not formally organized until the early spring of 1898, but Quealy and Kemmerer completed the initial planning and started selling lots in 1897. By August 1898, the name Kemmerer had been given to the new town. Kemmerer was incorporated on January 23, 1899.

Unlike the surrounding communities of Frontier and Diamondville, Kemmerer was founded as an independent coal town, which meant lots were sold, not leased from the coal company. Anyone who wanted a lot could own land and the house they built on it. Shortly after the incorporation of the town, several top officials from the Oregon Short Line Railroad Company joined with Quealy and Kemmerer in securing titles to additional land in the immediate vicinity of the town. They organized the Short Line Land and Improvement Company and dedicated the first addition to the town of Kemmerer in the spring of 1899.

Located in southwest Wyoming in Lincoln County along the Hams Fork River, Kemmerer became the county seat in 1911 when Uinta County, one of Wyoming's five original counties, was divided. The Hams Fork River that runs through Kemmerer was named after Zachariah Ham, a fur trapper who favored this area as part of the lucrative Green River drainage system.

After the town was founded, the Oregon Short Line Railroad began to construct large railroad yards, a four-stall roundhouse, and a depot to handle the coal shipments originating from the local coal mines.

The early ranchers who settled in the area before 1891 were able to obtain more than the 160 acres of land allowed under the Homestead Act of 1862. Most early ranchers raised cattle. The earliest written account of sheep in the Kemmerer area came in 1853, when Kit Carson drove more than 13,000 head of sheep from Fort Laramie to California. The sheep industries in Wyoming began to take shape between 1875 and 1885, starting a steady incline; but in recent years, the sheep industry has been hit hard as the numbers of sheep have declined and so have the sheep ranchers. The reasons have to do with predators, wool prices, and labor problems, some two-thirds of sheep men having shifted to cattle production or having sold out. There are approximately 50,000 head of sheep left in south Lincoln County.

Cattle ranches in the area started with the first settlers in the 1880s. This industry has continued to increase over the years. There are approximately 100,000 head of cattle in south Lincoln County. On May 27, 1897, Patrick Quealy organized the Hams Fork Cattle Company. The formation of the cattle company was not particularly significant at the time, but Quealy was convinced that he had found the ideal way to incorporate the cattle, coal, mercantile, and land companies, all of which would complement each other. The date this company merged with other Quealy and Kemmerer companies is unknown.

Cattle and sheep were an important part of the company stores. They had their own slaughterhouse and butchered the meat as it was needed. No inspections were necessary at the time. There were also no laws against killing wild game and selling it.

Wild meat was on the menus of the diners, and when dignitaries would come into the area, it was not uncommon for them to wire ahead to make sure they would have an elk steak for dinner. Elk meat, along with cattle and sheep, would be shipped on the train back east for sale.

The area's most famous cattle rancher was a lady cattle rustler named Anna Richey, also called Annie. Known as the Petticoat Rustler, she was the only woman in Wyoming to be convicted of cattle rustling. In 1919, after Richey's cattle were shipped by train to the stockyards in Omaha, Nebraska, it was discovered that 32 head of cattle that were shipped had been rebranded with Richey's brand. She was convicted of cattle rustling on November 25, 1919; the case was appealed to the Wyoming Supreme Court, and it took two years for an answer to the appeal. The Wyoming Supreme Court upheld the sentence imposed by the district court, one to six years in the penitentiary at Canon City, Colorado. In May 1922, shortly before she was to begin her sentence, Richey was poisoned. It remains Kemmerer's unsolved mystery.

Education in Wyoming, on a structured scale, started in 1852 when the first school was opened at Fort Laramie. As the population of Kemmerer began to increase, it became necessary to provide a schoolhouse and teachers. Kemmerer's first school opened its doors on January 11, 1899, in the Quealy home with 45 students. School was held at this location for approximately a year and a half until a new school could be built. The new school opened in the fall of 1902 with 100 students. The first teacher to the Kemmerer area was Grace King. Because the new school did not have seating for all the students, a larger school was constructed. This new school, opened in 1910, was called the Kemmerer Grade School or Central Grades, and it accommodated all age levels. At this point, most students were young. High school students were mostly girls, as many of the young men were working in the coal mines. This building eventually became too small and was replaced in the mid-1920s with a larger brick building, and that building was vacated in the early 1970s when the current elementary school was built. A high school was built next to the grade school in 1913, and that building became the junior high school after the completion of the present high school. A new junior high school was added on to the high school in 1968.

On April 14, 1902, retail giant James Cash Penney started his large chain of clothing stores in Kemmerer. The first store was named the Golden Rule. Most of the people in the area were coal miners who shopped at the company store; many doubted that the Penney store would succeed. But on the opening day of the store, it made $466 and, at the end of the first year, profited over $29,000. By August 1904, the store had grown and was moved to a larger location. That same year, two additional stores were opened, one in Rock Springs, Wyoming, and one in the coal camp of Cumberland, Wyoming, just south of Kemmerer. In 1929, the Kemmerer store moved to its present location on J. C. Penney Drive. On January 17, 1913, the Golden Rule stores were incorporated as the J. C. Penney Company. As the number of stores increased, the company headquarters moved first to Salt Lake City and, in 1914, to New York City. The company offices are currently located in Texas. James Cash Penney died on February 12, 1971, at the age of 95. He is buried at Woodlawn Cemetery in New York City.

With the passing of the 18th Amendment to the Constitution, prohibiting alcohol, in 1919 and the passing of the Volstead Act, prohibiting wine, in 1925, nationwide fame came to the area. On June 30, 1919, a crowd gathered to mourn the passing of "John Barleycorn." At midnight, the saloon doors were closed, and preparations were made to change the saloons into legitimate businesses.

However, the law didn't discourage people who wanted to drink or those who supplied it. Kemmerer and Diamondville became national centers for bootlegging and major suppliers of liquor for Chicago and the Midwest. "Kemmerer Moon" was a favorite of the speakeasies. This fame gave Kemmerer the nickname "Little Chicago."

The Depression was difficult for people all over the United States, but more so in the mining camps where coal orders were at a low point. Most mines were working only one day a week. This was not enough money to keep most families going, so some of the miners turned to bootlegging. There were a few legal purposes for liquor, mainly medicinal. During this time, whiskey was thought to be the best cure for what ails you. This all ended in 1933 when the 21st Amendment to the Constitution was passed, repealing Prohibition. As a result of a resolution passed by the legislature, Wyoming could not vote on the 21st Amendment until 1934. On April 15, 1935, Kemmerer received the first legal shipment of liquor.

Although Kemmerer has continued to grow and change over the past 111 years, it has remained a coal mining community with a strong sense of pride. The Kemmerer area is dependent on fossil fuel–based industries for its livelihood. The gas and oil fields, coal and trona mining, and coal-fired power plants are an ever expanding field of exploration that continues to provide employment for the area. The fossils that are found in the area have also given Kemmerer worldwide recognition, as Kemmerer is known as the "Fossil Fish Capital of the World."

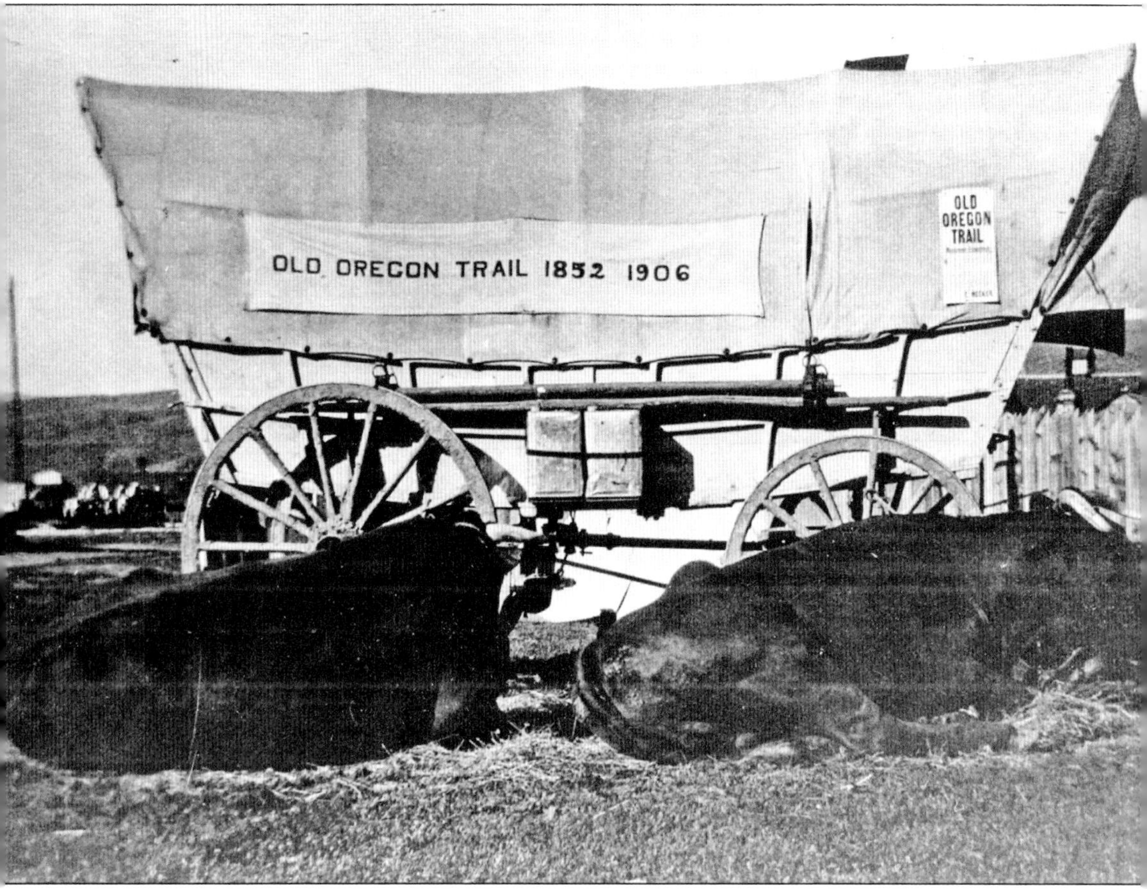

The Kemmerer area and the Hams Fork Valley have the distinction of being one of the areas where the emigrants made their roads traveling from east to west, most travelers going to Oregon. Many emigrant journals mention the Native Americans in the area and the rough terrain they had to travel over. Many stated that the surroundings were beautiful, but the environment was unforgiving.

One

Beginnings

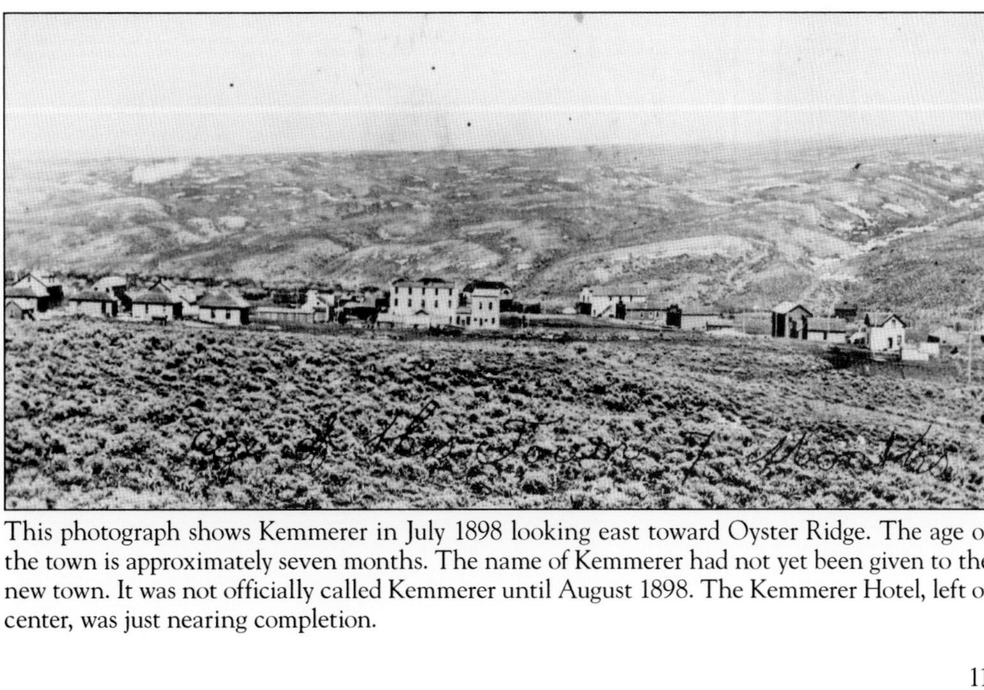

This photograph shows Kemmerer in July 1898 looking east toward Oyster Ridge. The age of the town is approximately seven months. The name of Kemmerer had not yet been given to the new town. It was not officially called Kemmerer until August 1898. The Kemmerer Hotel, left of center, was just nearing completion.

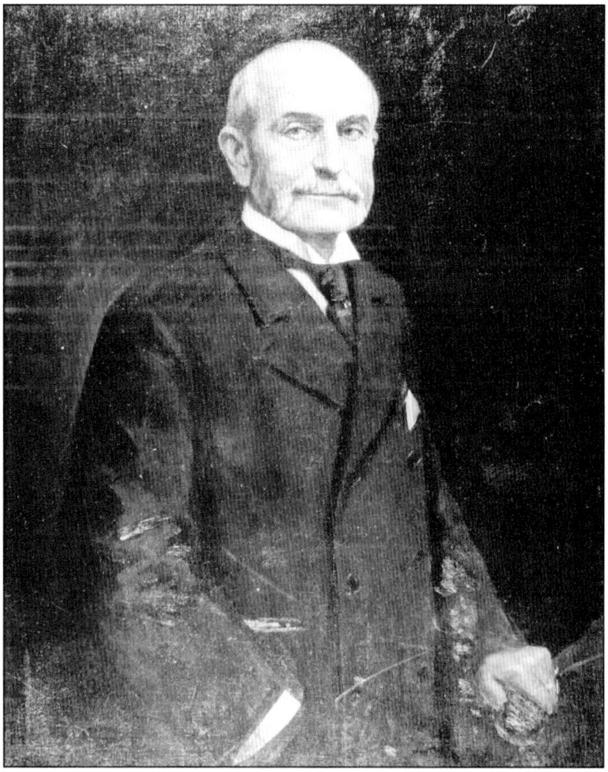

The two founding fathers of Kemmerer were Mahlon S. Kemmerer (left), for whom the town is named, and Patrick J. Quealy (below). These men charted the path and led the way in the development of the coal mining industry in this area. The two men formed the Kemmerer Coal Company and the company town of Frontier just north of Kemmerer. Quealy arrived in the area in March 1881 while he was investigating coal prospects for the Union Pacific Railroad. He realized that the area held large coal deposits, so he started to acquire lands along the railroad right-of-way. In 1895, Quealy traveled to New York City to find investors for his Wyoming coal holdings. At this time, he was introduced to Kemmerer, which resulted in a lasting partnership.

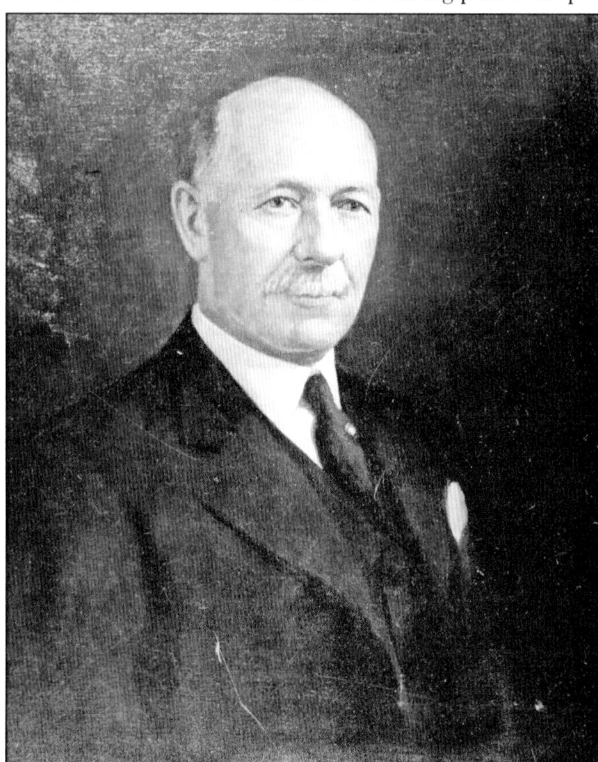

This photograph was taken in Quealy's yard shortly before Kemmerer's death in 1923. The two men were not only successful businessmen but also good friends. Kemmerer remained the head of the Kemmerer Coal Company empire until his death. Even though the Kemmerer family did not reside in the area as the Quealys had, they were always very proud of the community.

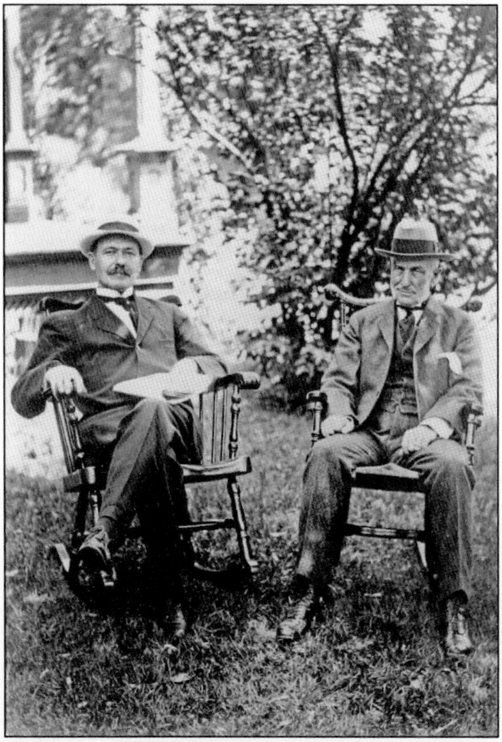

The need for a bank had been contemplated for a few months by Quealy. After rumors about a bank being built in Diamondville by Anaconda Copper Company, the competing coal mine, Kemmerer sent orders to Quealy to begin construction of a bank. The First National Bank opened for business in July 1900. Total deposits in the first year were approximately $145,000.

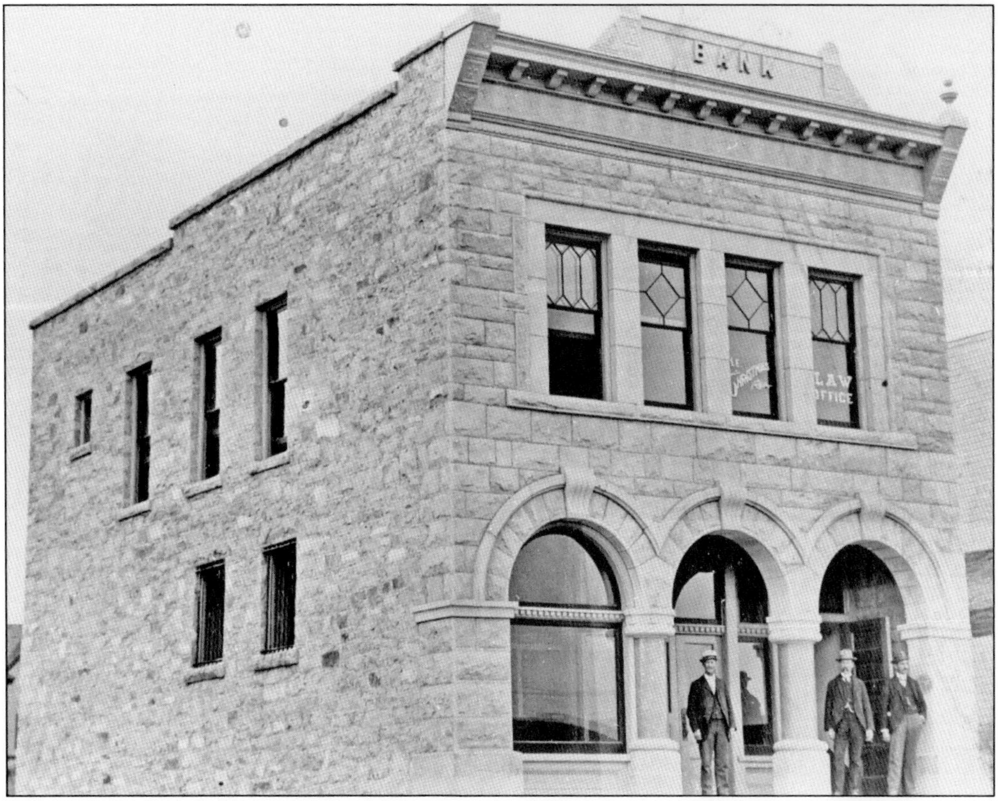

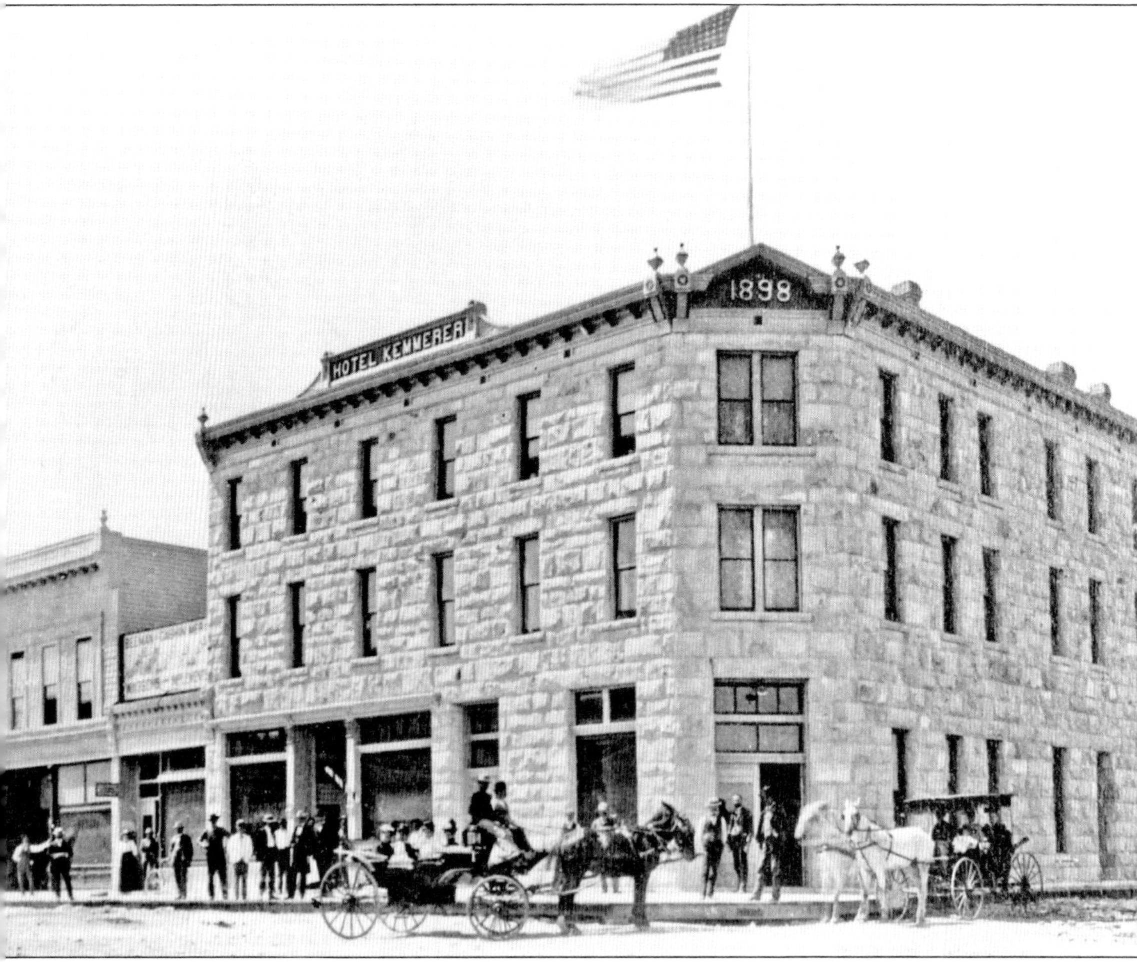

Construction of the Kemmerer Hotel began in 1897, but there were several delays caused by financial problems and the cold winter weather. The hotel was completed and opened for business in August 1898 under the ownership of Robert White. Native stone used to build the hotel was cut from the nearby stone quarry located in Oakley, just south of Kemmerer. The hotel was the first large hotel to be built in the new community. Shortly after the town of Kemmerer was incorporated and a town council appointed, the first town meeting was held in the hotel on February 28, 1899. The hotel was able to weather economic booms and busts. Livestock and coal production, the foundation of Kemmerer's economy, provided enough stability for the hotel to survive. This gave her the nickname the "Grand Old Lady."

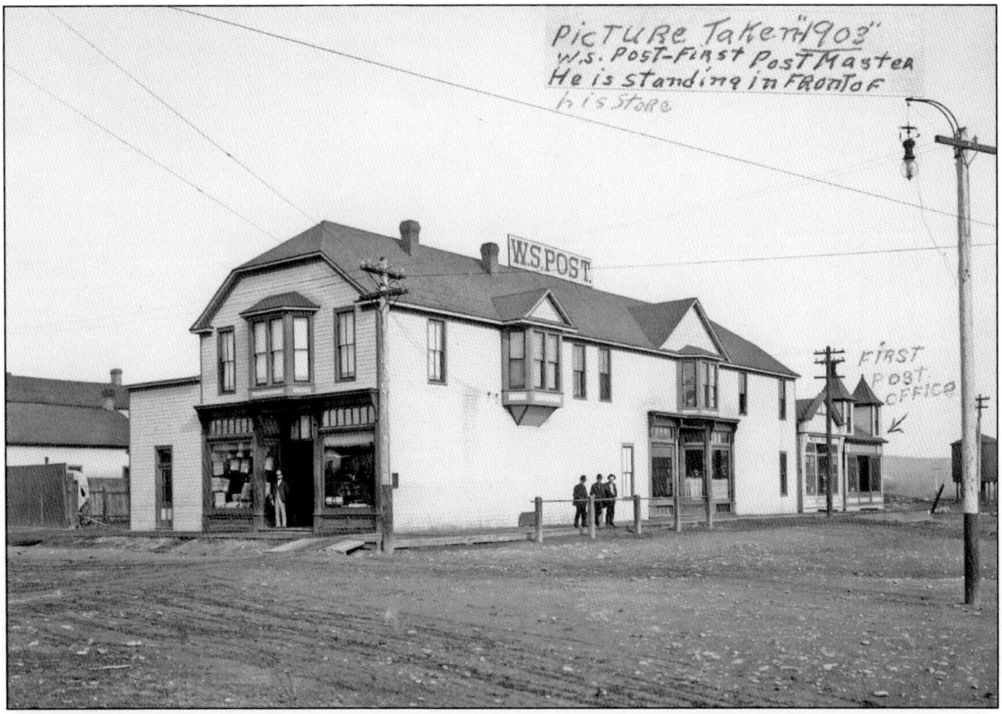

The beginning of a post office in the area started in the old town of Hams Fork, west of Kemmerer. In 1890, W. S. Post, a telegrapher for the Oregon Short Line, ran the post office for the Hams Fork train depot out of his general store. In 1897, Post moved the store to Kemmerer. The first Kemmerer Post Office was established in the Post building on December 1, 1897, with William S. Post as postmaster. The post office was eventually relocated to a building next to the Kemmerer Hotel. The location served both as post office and stagecoach stop for the stage running between Kemmerer and Big Piney. Kemmerer has had mail delivered to homes since 1937.

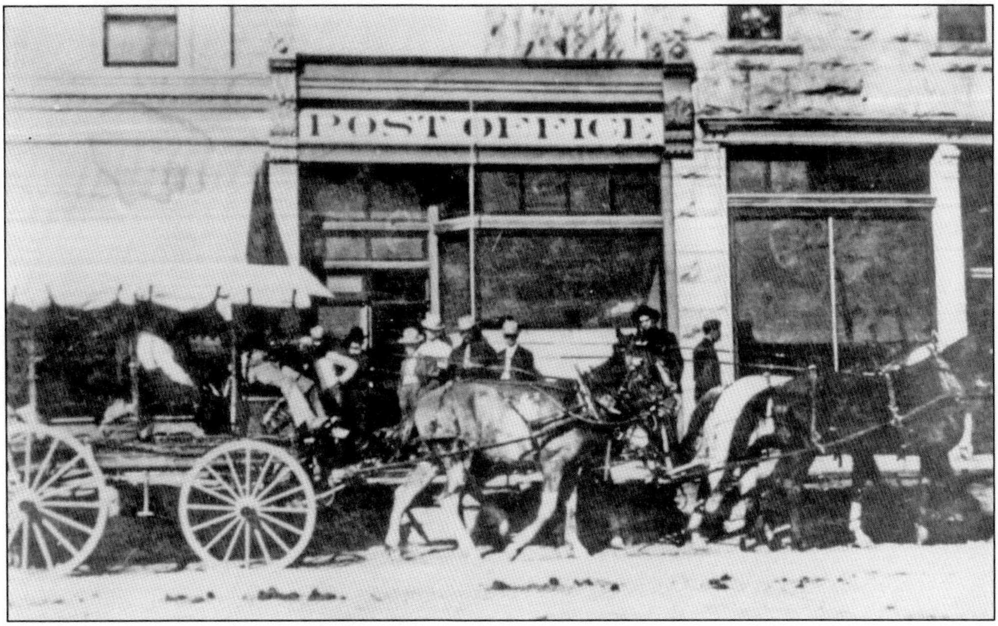

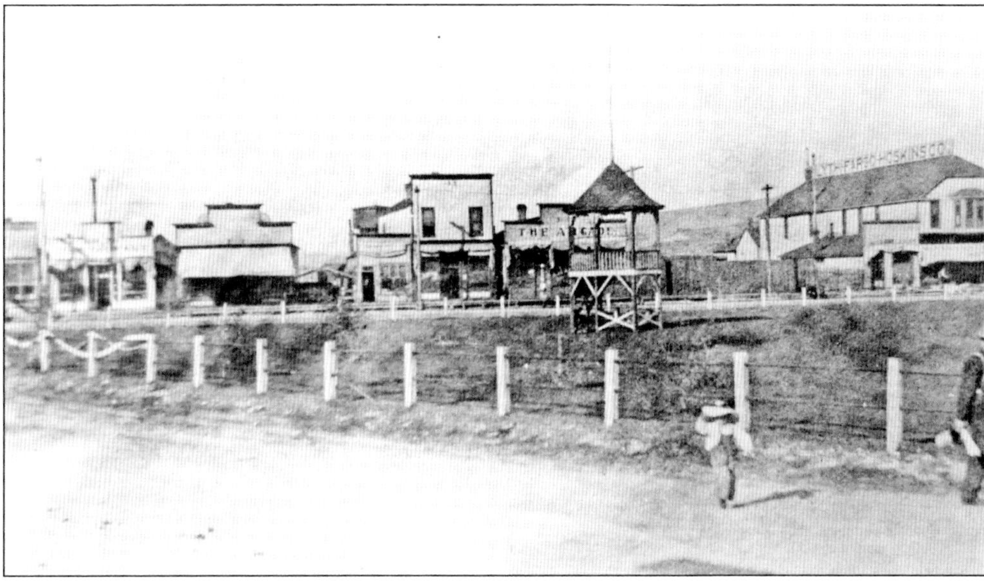

It is easy to see in these two photographs the rapid growth of Kemmerer. From the beginning in 1897 to 1903, the town buildings and population increased dramatically. Because Quealy and Kemmerer wanted to attract the finest to Kemmerer, they understood that the first impression a person got of a place was always a lasting one. Most of the first buildings in town were large, tall buildings. By 1903, the downtown area had a bank, a hotel, an opera house, and several stores, including the J. C. Penney store. It is suggested that Quealy encouraged and sometimes helped financially to get these first buildings erected, knowing they would draw attention and this would lead to the eventual growth of Kemmerer. It was with pride that the city fathers and those first merchants advertised their new town.

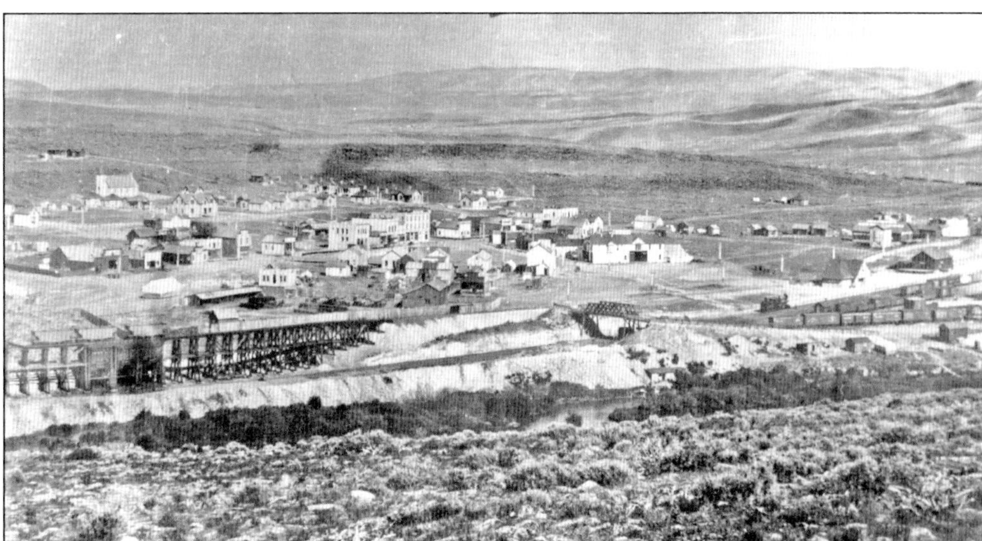

Two

The Kemmerer Coal Company and Frontier

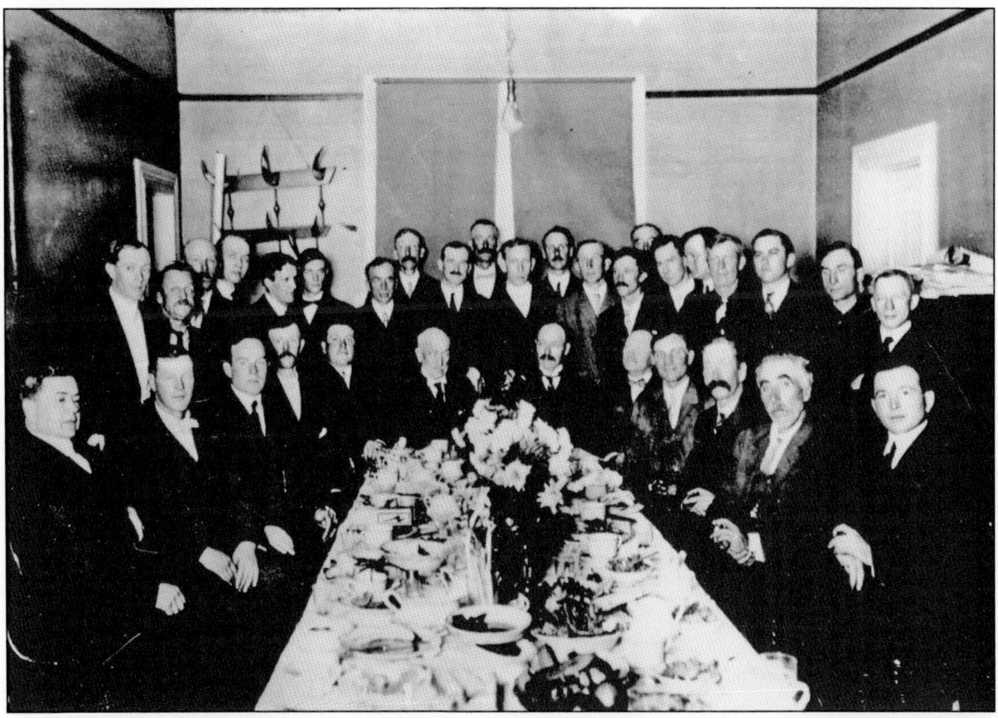

The first officers of the newly formed Kemmerer Coal Company, Kemmerer (left) and Quealy (right), are seated at the head of the table. The Kemmerer Coal Company was incorporated in May 1897, and the incorporation papers stated that the company was located in Kemmerer, Uinta County, Wyoming. This was a mistake since at this time the town of Kemmerer did not exist. The company headquarters were located in Frontier.

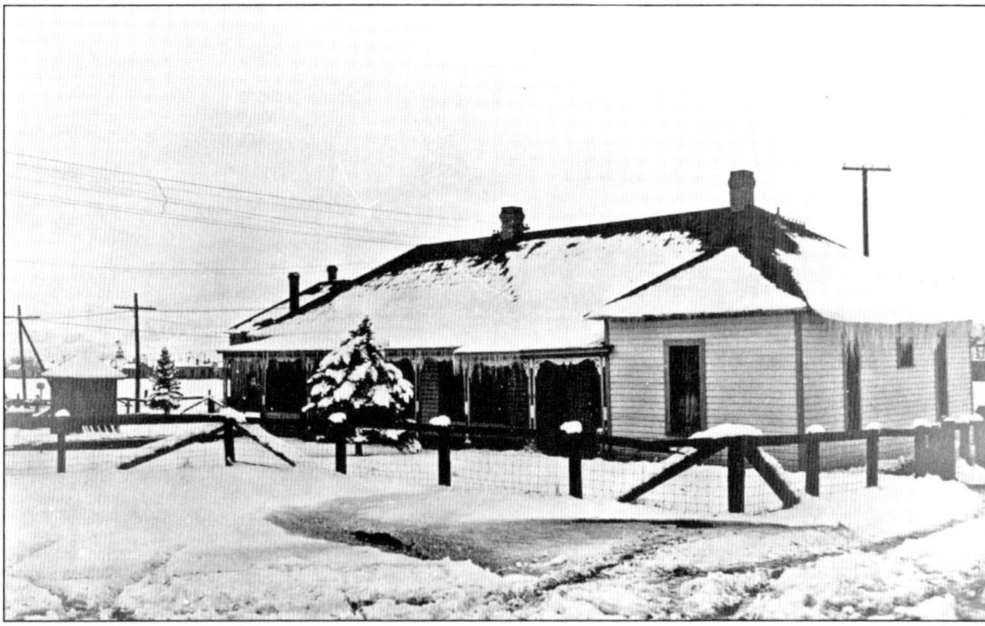

These photographs show the first offices of the Kemmerer Coal Company, located in Frontier, Wyoming, and the first office staff of the company in 1897. The house was used as offices and also the Quealy home until their home was completed in Kemmerer. The Quealy home in Kemmerer was also used as a school. After the company built a bigger office, the Kemmerer family would stay in the house when they would visit. The office staff managed the mine operations, mine payroll, and the accounts in the company store. They reported to Quealy and Kemmerer on a weekly basis.

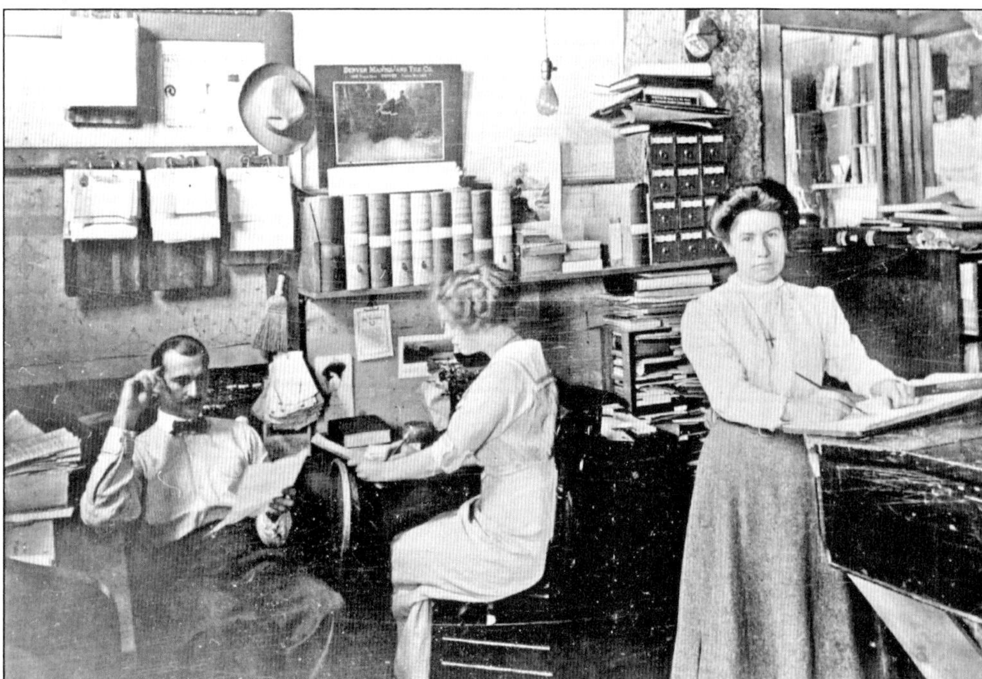

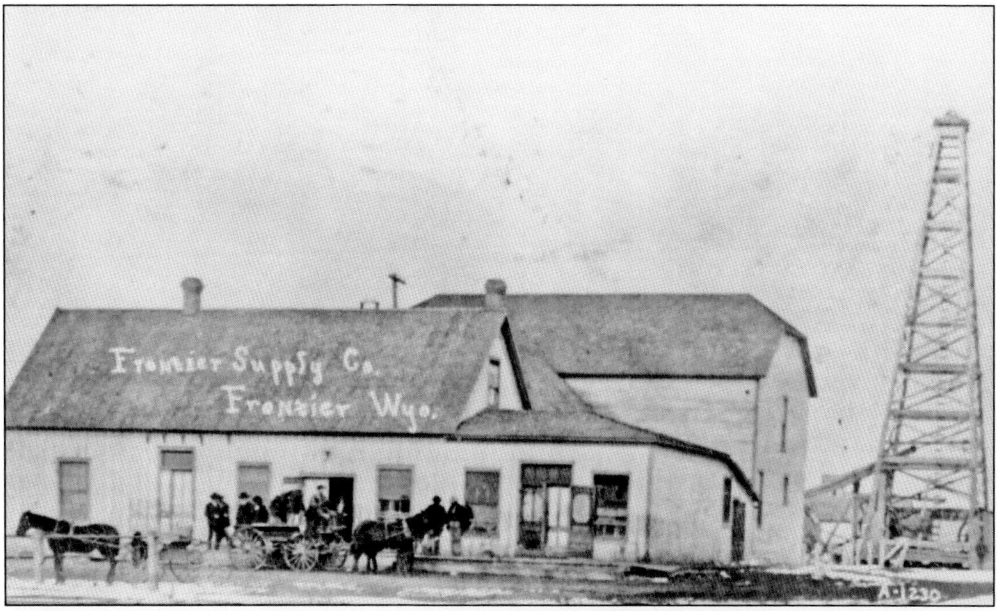

The Frontier Supply Company was organized on May 25, 1897, by Quealy and Kemmerer as part of the Kemmerer Coal Company. A store was also built in Kemmerer after the town of Kemmerer was founded. Both of the stores were company owned; they handled groceries, meats, clothing, shoes, dry goods, hardware, lumber, hay, and feed for livestock. The company store used coupons or scrip, which was issued against the miners' pay, and the men with large families normally got a receipt or statement instead of a paycheck. This is what is meant by the saying, "I owe my soul to the company store." The company store also contained a post office, which was officially opened on January 20, 1899, with William S. Avery as postmaster and general manager of the store.

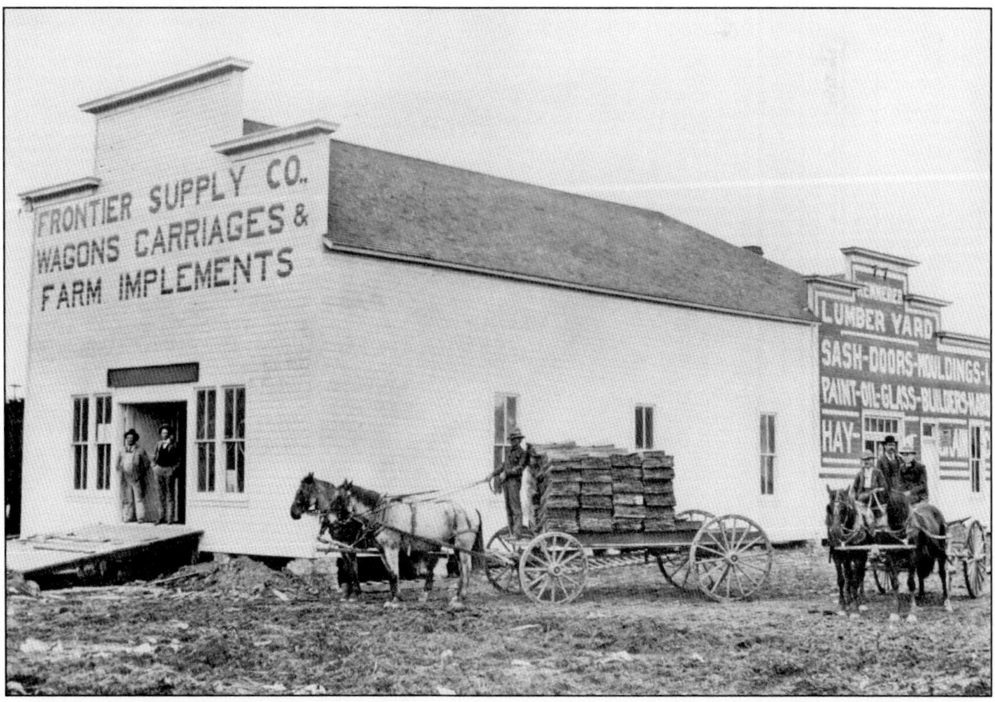

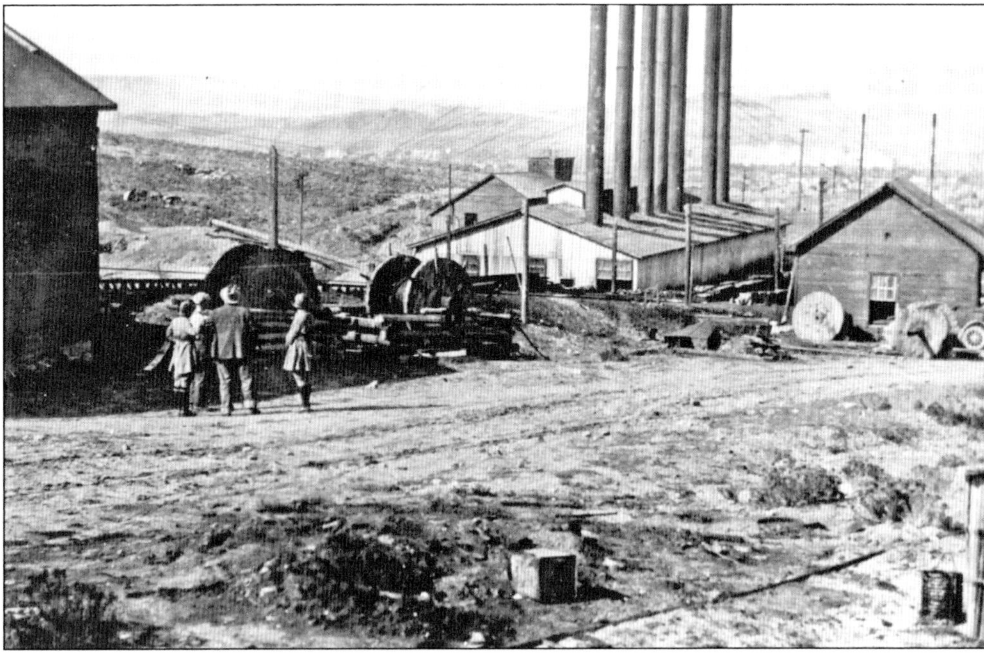

In July 1897, preparations began to start digging the first mine of the Kemmerer Coal Company; the mine would be called the Frontier No. 1. The mine was located east of Frontier on Oyster Ridge. Quealy had some of his old mining equipment, which had been in storage after his Rock Springs mines were sold, shipped to Kemmerer. Most of the heavy equipment, however, came from back east. Now that the equipment was in place, mining could begin. Quealy and Kemmerer were on hand on October 5, 1897, to witness the first shipment of coal. In its first year, some 24,940 tons were taken from the new mine. By December, the new coal company was well underway.

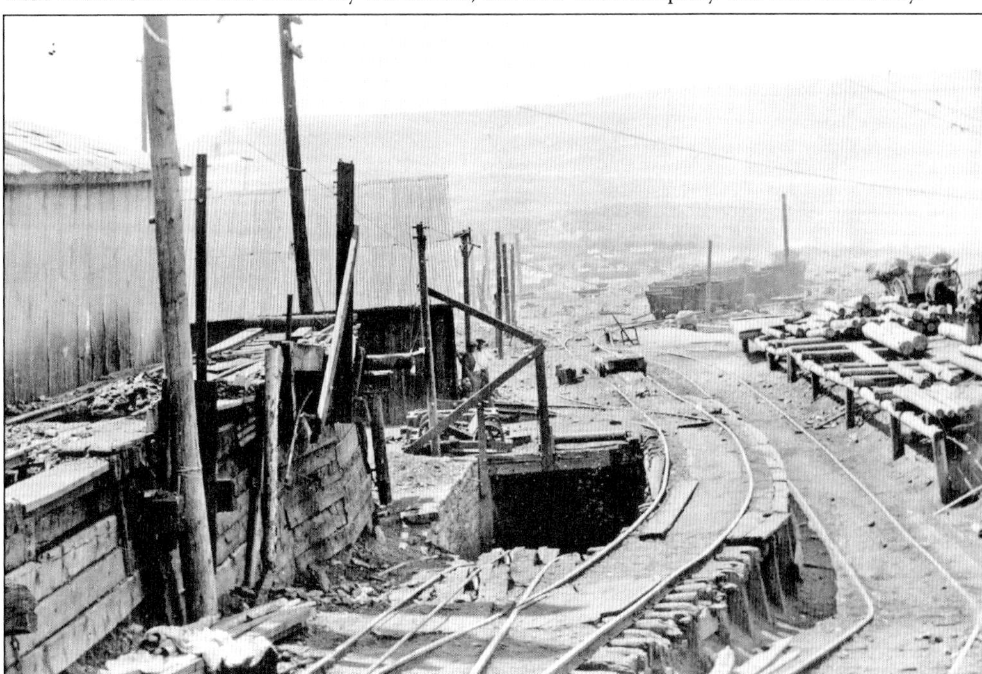

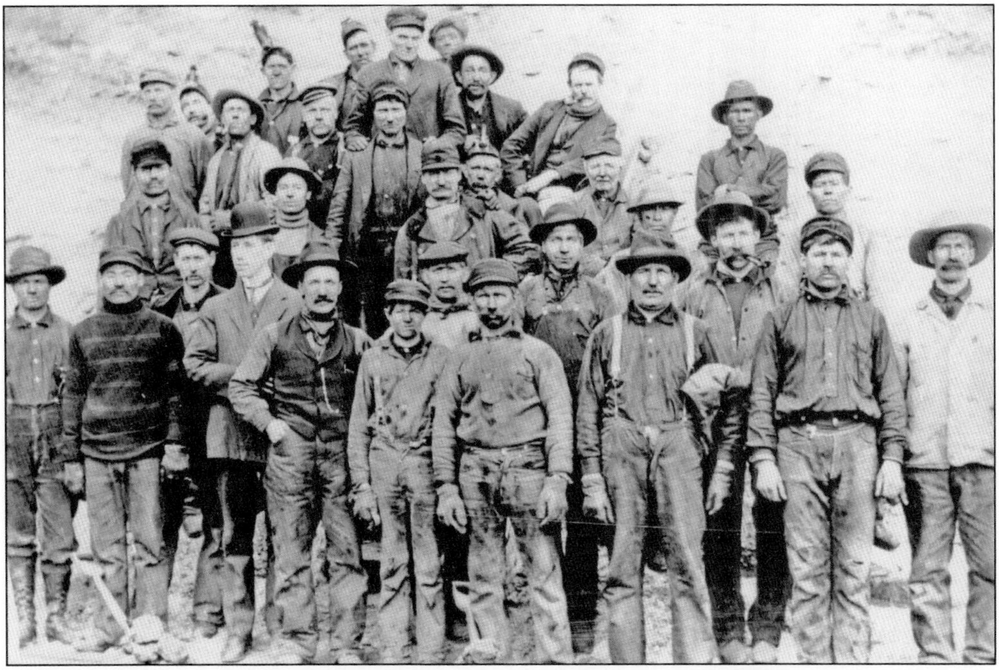

The first coal miners entered the new Frontier No. 1 mine in 1897. The Frontier mine was in operation from 1897 until 1926. Large numbers of immigrants came to the area for a better life. Underground coal mining was a dangerous job, and the miners knew the possibility for mining accidents. A normal day for the miners lasted from daylight until dark, spending the entire day underground; the only light would be from the lamps on their hats or the ones hanging on the walls. The men would work 12-hour days for very little pay. The Western Federation of Miners was established in Diamondville in March 1898, but this union was not allowed to organize a local in Frontier until 1903. This helped change the workday to 8 hours.

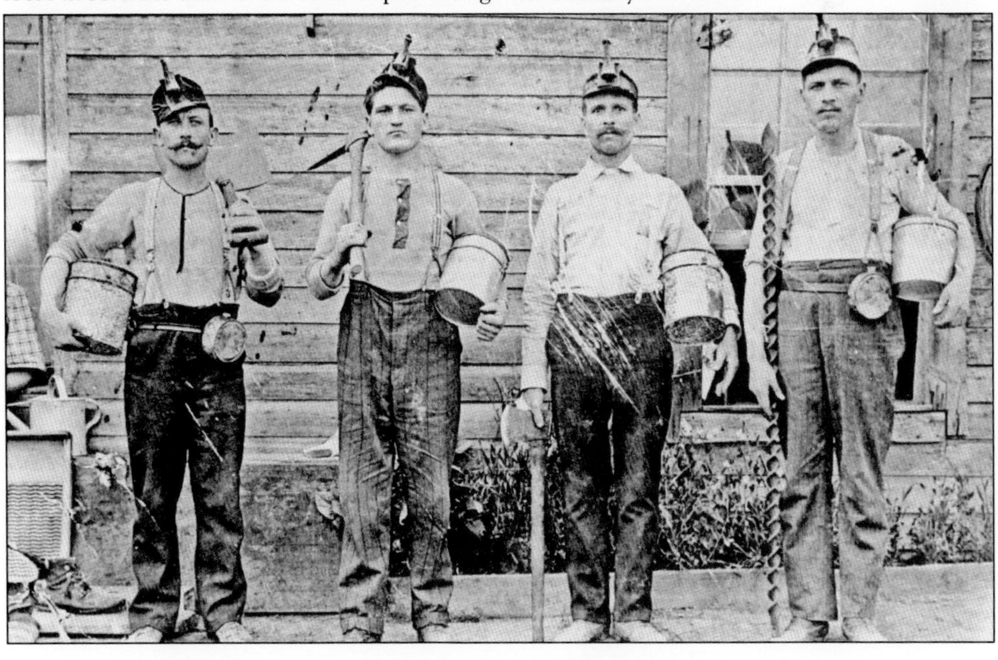

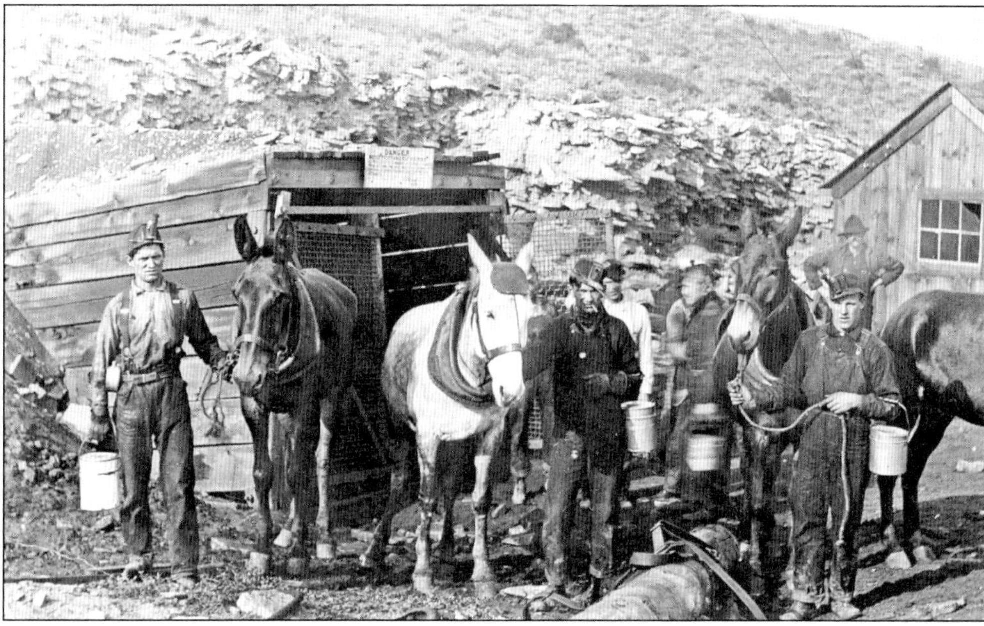

Before the rails were built in the mines, the loaded coal cars were pulled to the surface with horses or mules. The miners who took care of the animals underground were called mule drivers. After every shift, the animals were brought to the surface and placed in the company barns. The Kemmerer Coal Company mines were the only area mines to do this. The animals that stayed underground for long periods of time would usually go blind. The young men who were too young to work underground would be hired by the company to take care of the company barns and animals. They would rub down and feed the horses at night and have them ready in the morning to go to work.

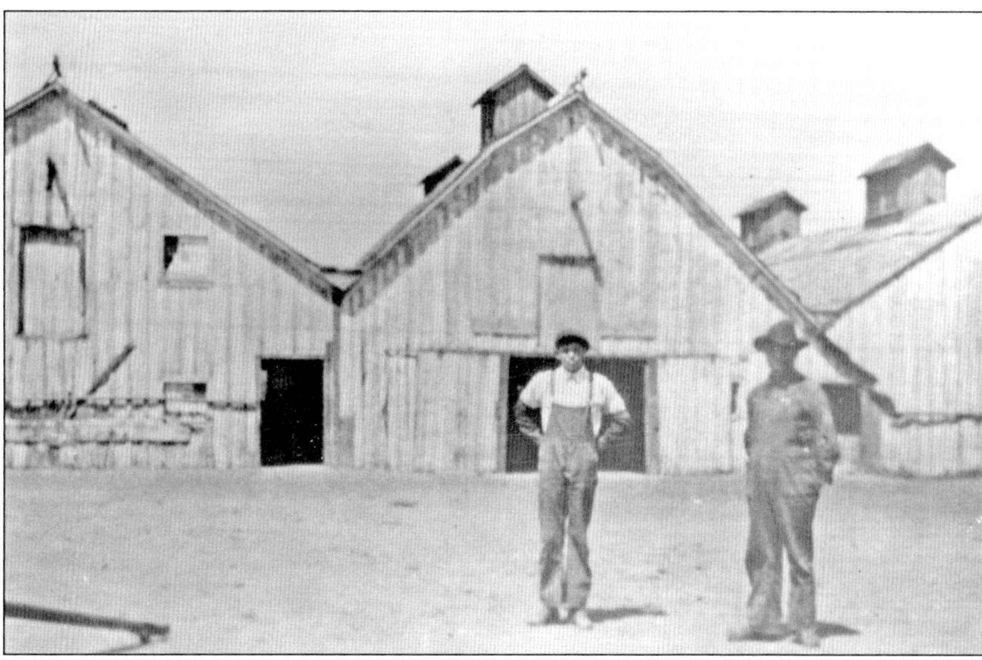

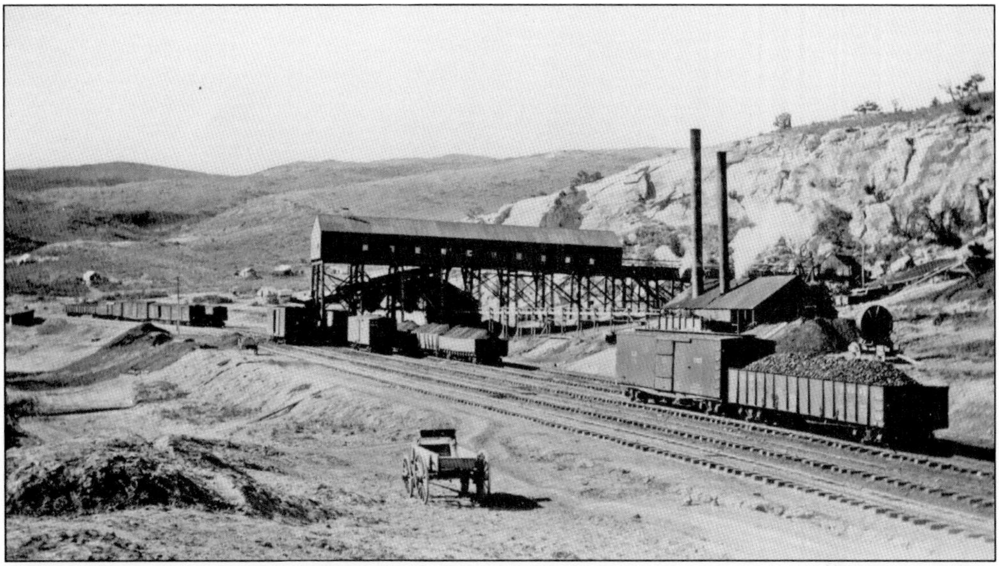

During the 84-year history of the Kemmerer Coal Company, it has owned and operated 10 mines. The last underground mine owned by the company was the Brilliant Mine, which closed in 1964. The last Kemmerer Coal Company mine still in operation is the Elkol Mine. It opened in 1908 under the name of the Elk Coal Company, and several companies owned the mine until Kemmerer Coal acquired it in 1920. In 1950, underground mining was suspended at this mine, and surface mining began. This was a more cost effective way to mine coal. The Kemmerer Coal Company owned the Elkol Mine until 1981, when it was sold to Pittsburgh and Midway Coal Company. It is now the deepest open-pit coal mine in the United States.

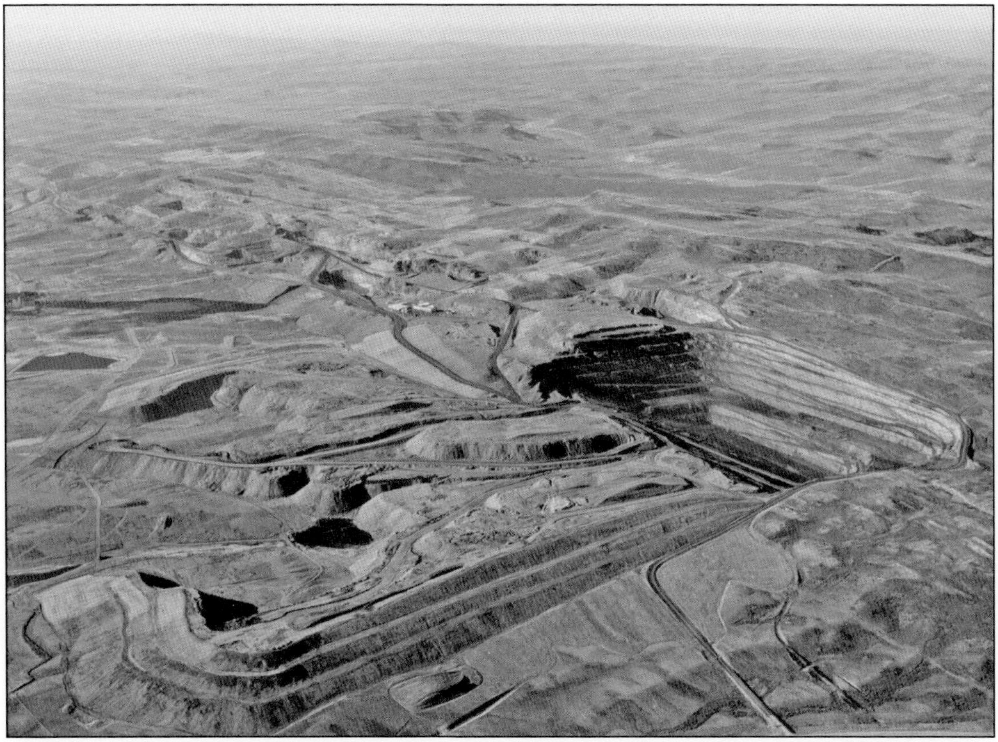

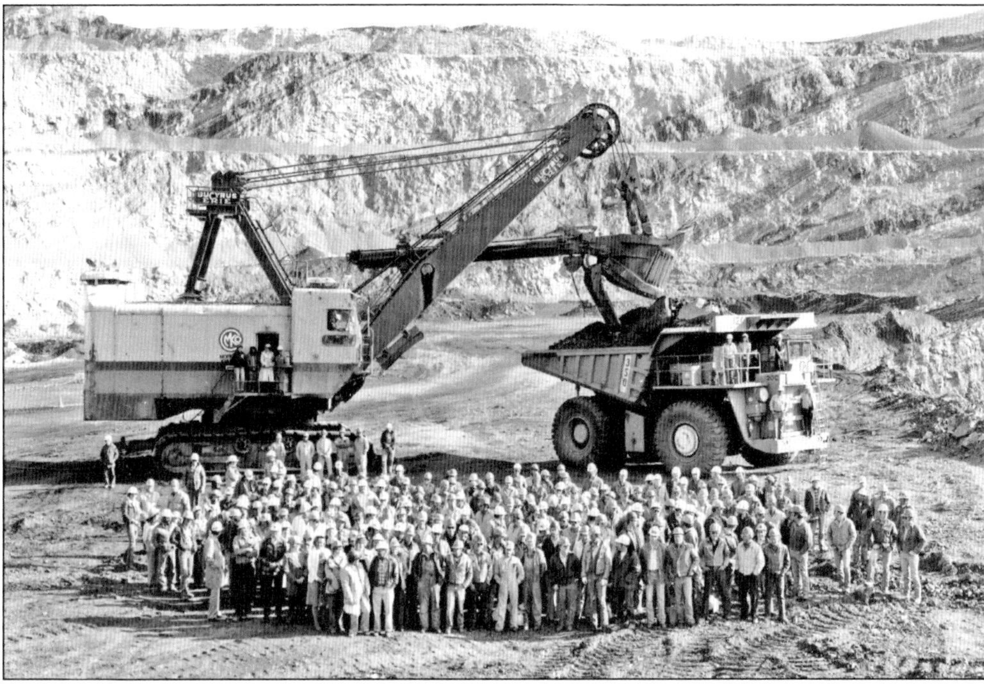

The photograph shows men and women coal miners in the pit of the Elkol Mine. Women did not work in the mines until 1943, when there was a shortage of manpower because of the war. Women were then hired to work the tipples since it was considered bad luck to let women underground. The Kemmerer Coal Company paid the women the same wages as the men.

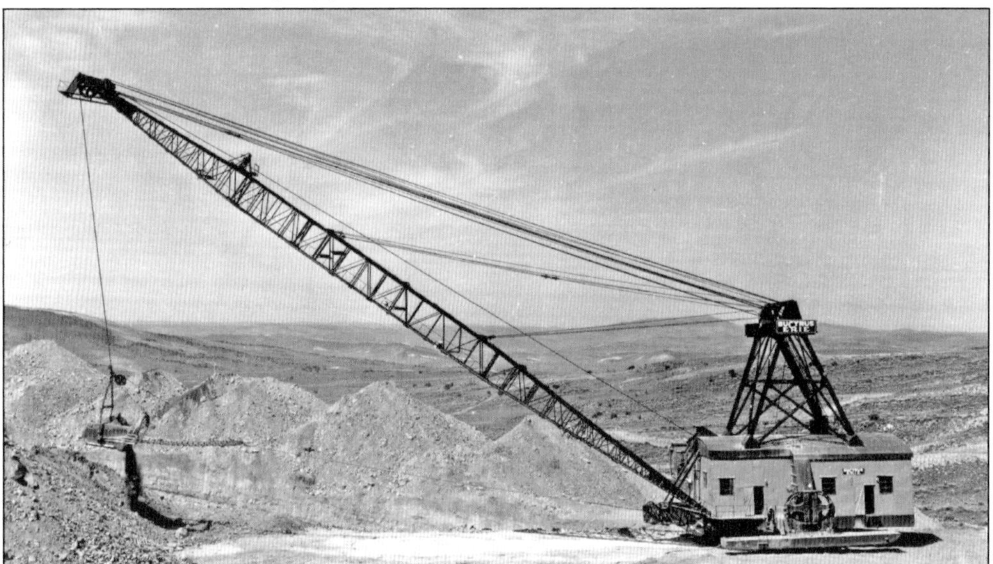

The Monigan Dragline, known officially as the Bacyrus Erie 480 W Dragline, was used to move overburden. The Monigan was purchased by the Morrison-Knudsen Company in 1961; this company leased the mine from the Kemmerer Coal Company. The Monigan was dismantled and sold in late 1984. The miners who worked on the Monigan stated that "it was slower than a turtle and bigger than a dinosaur."

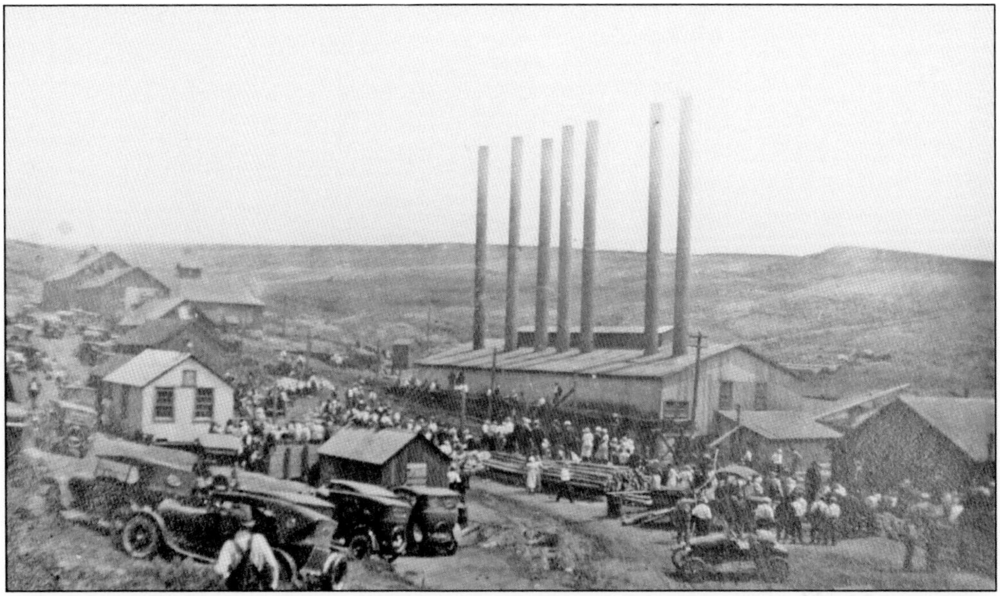

Wyoming's second worst mining disaster occurred at the Kemmerer Coal Company's Frontier No. 1 mine on August 23, 1923. Ninety-nine men were killed by a methane gas explosion and fire. Thirty-seven men were rescued after being trapped for several hours. Some of the miners had taken the day off as a religious holiday, or the loss of life would have been far greater. The accident touched the lives of everyone in the communities of Frontier and Kemmerer. Families rushed to the mine entrance waiting to hear some news of their loved ones. The inquest that followed the explosion concluded that the accident was caused when a fire boss attempted to relight his safety lamp with a match, which ignited the methane gas that had accumulated in the mine. The fire boss, Thomas Roberts, was the last body recovered.

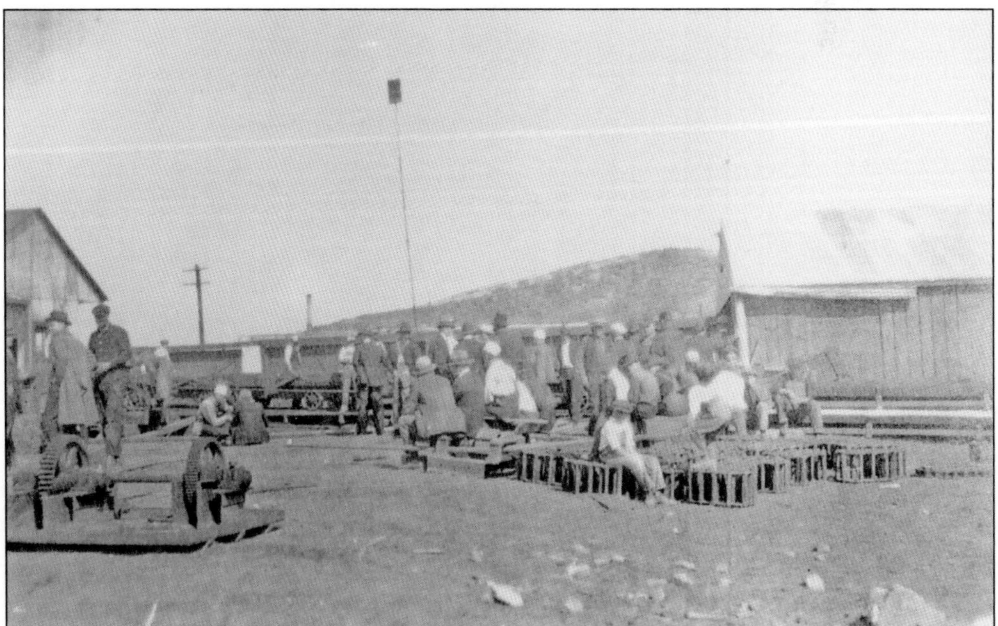

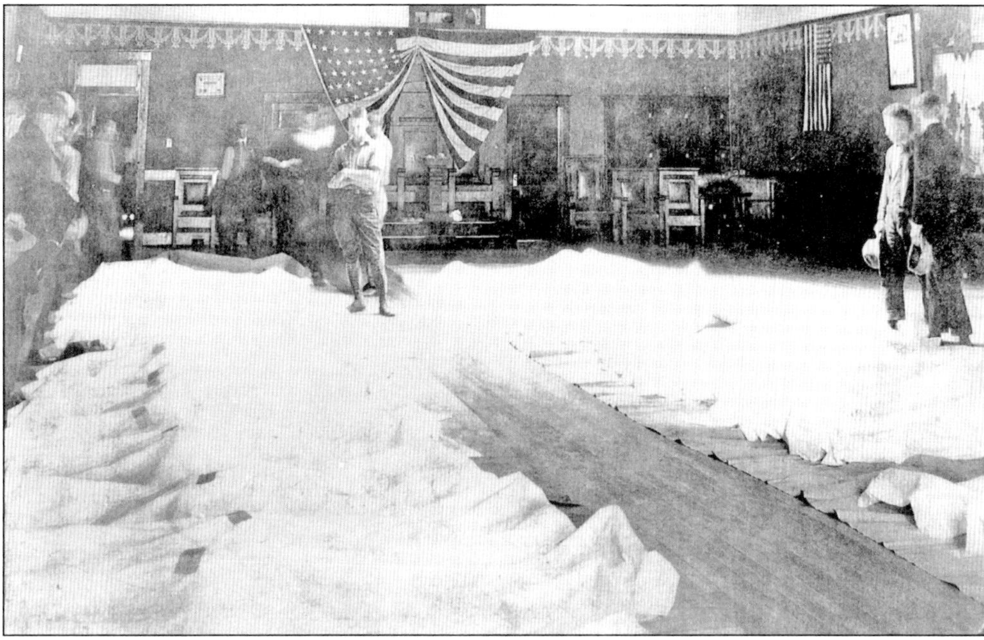

Soon after the disaster, preparations were made to take the dead to the International Order of Odd Fellows Hall in Kemmerer, where members of the IOOF and volunteer workers started the grim task of arranging and identifying the victims. The miners from the Cumberland mine started digging graves at the Kemmerer Cemetery; seven mass graves were dug. Preparations were also being made in the Triangle Park, where a platform was built for the speakers. The caskets were placed around the Triangle for public mourning. Hundreds of people from all the surrounding communities attended the services. At the cemetery, every miner was given a 21-gun salute and "Taps" was played. The men who gave their lives in the mines prompted the workplace safety standards in place today.

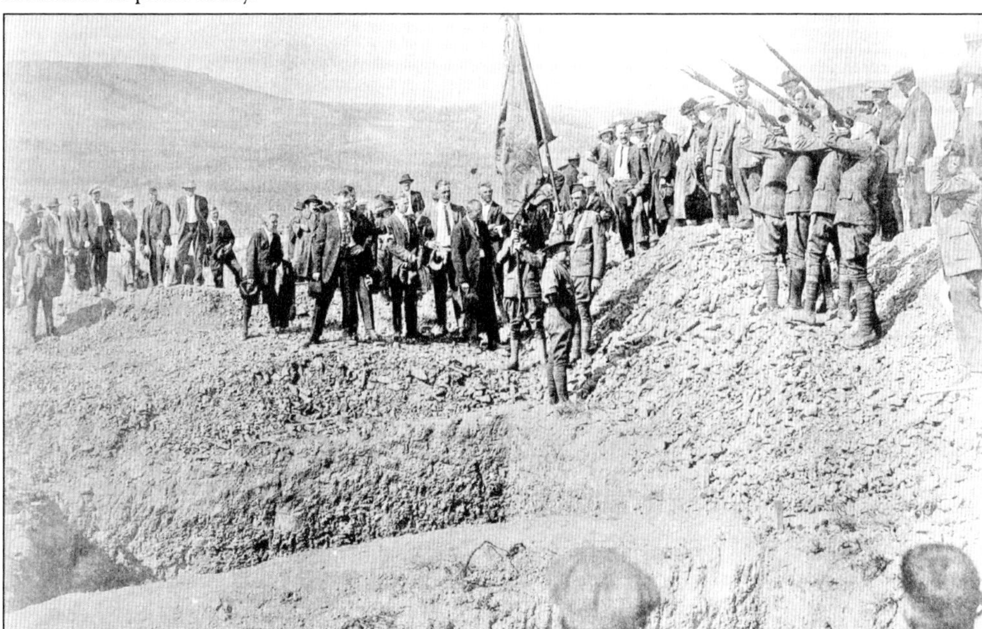

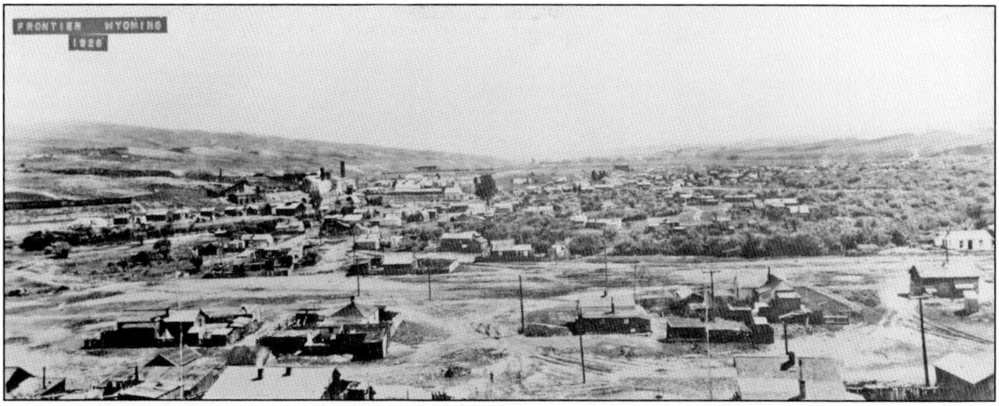

Frontier, started in 1897, was the coal camp for the Kemmerer Coal Company. The Hams Fork River divided the camp; the residents on the north side of the river would use a foot bridge built in 1914 to get to the company store, post office, and the Kemmerer Coal Company offices. The bridge was affectionately referred to as the swinging bridge. The coal company would not sell the land, but it did build tenement houses for the miners to rent for as long as they were employed by the company. By 1909, there were not enough tenement houses for all the miners, so the company allowed the miners to rent ground and build their own house. By 1926, the population of Frontier was over 2,000, and Frontier had changed from being a coal camp to a town.

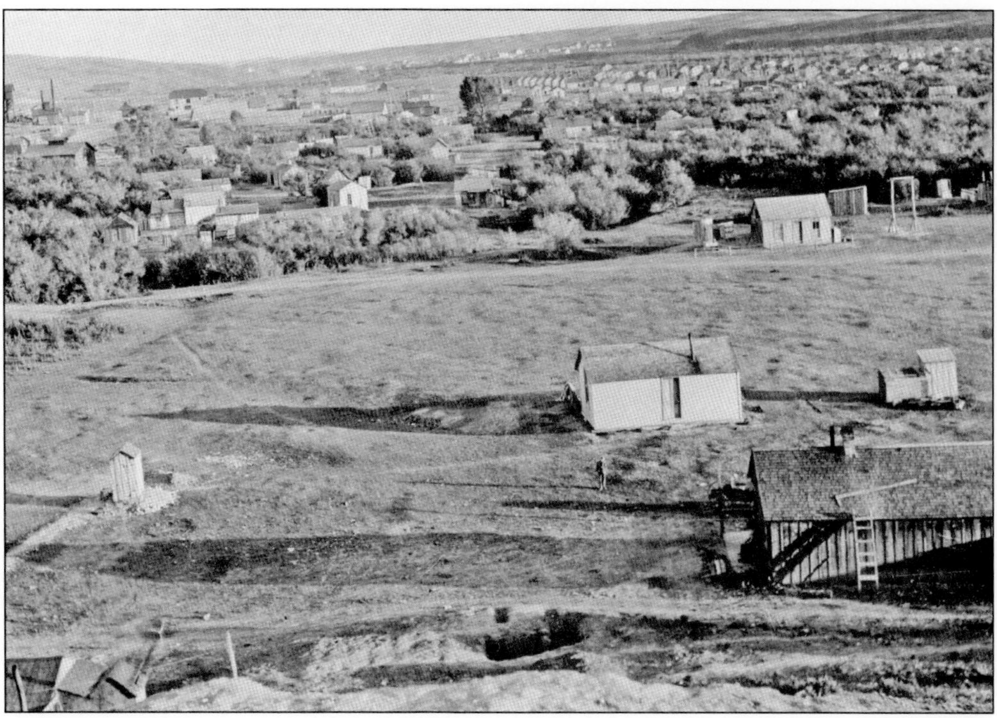

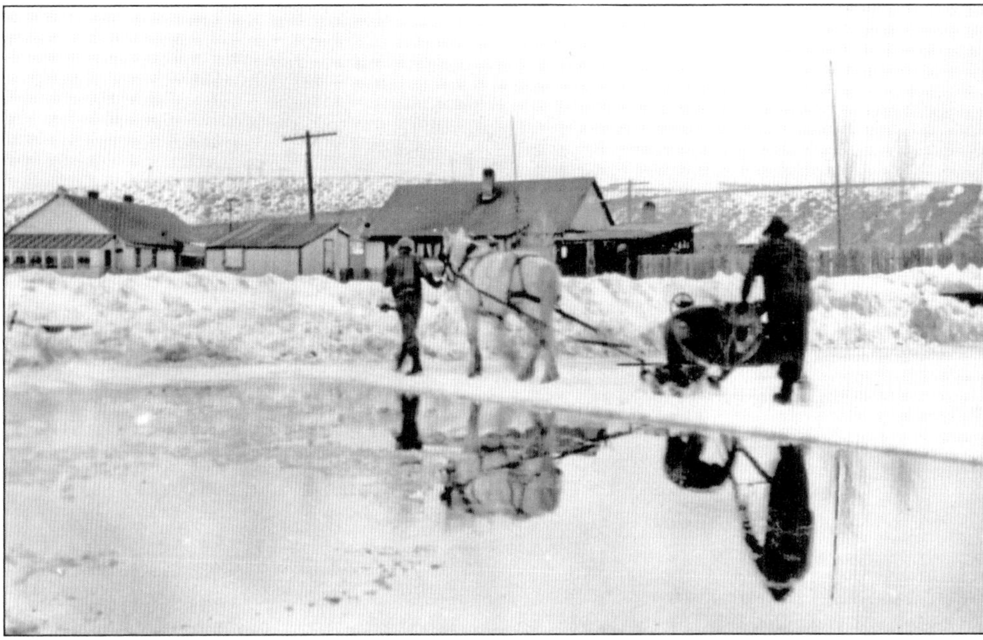

Ice harvesting was needed to cool the first home refrigerators and to cool the meatpacking houses. Most towns had an icehouse to store the ice. The buildings had double walls, and sawdust from the company sawmill was used for insulation. When ice cutting began in this area, it was sawed by hand, but later a Model T Ford engine was used to power the saw. Just north of Frontier on the Hams Fork River was the ideal place to cut ice. The ice was cut into large blocks and pulled from the river with large tongs. A horse would pull the blocks onto a platform and into a wagon, and later a truck was used for ice delivery.

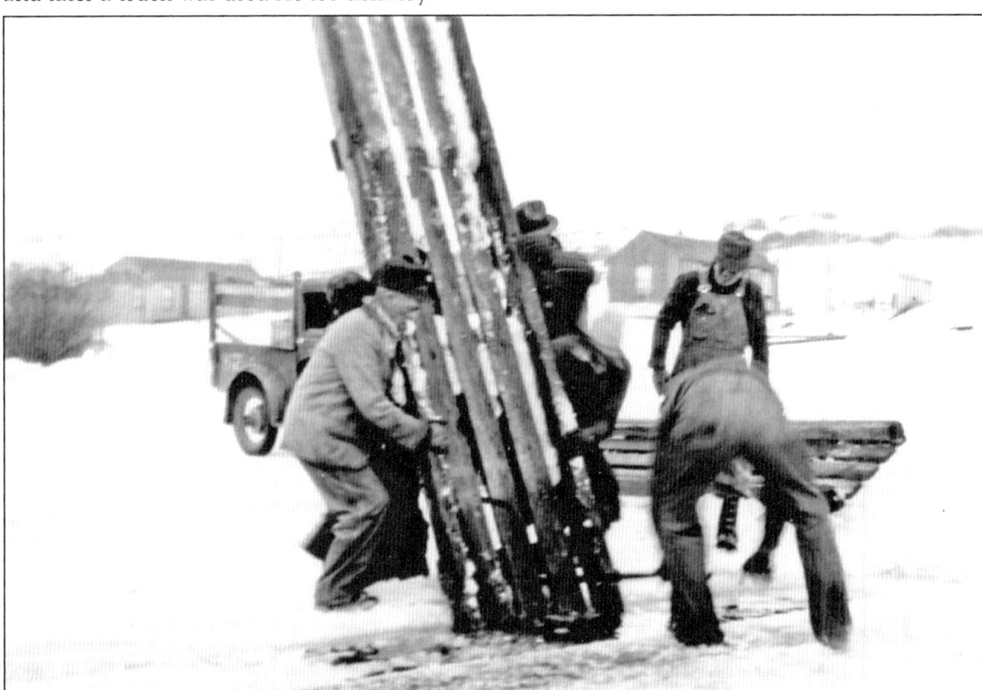

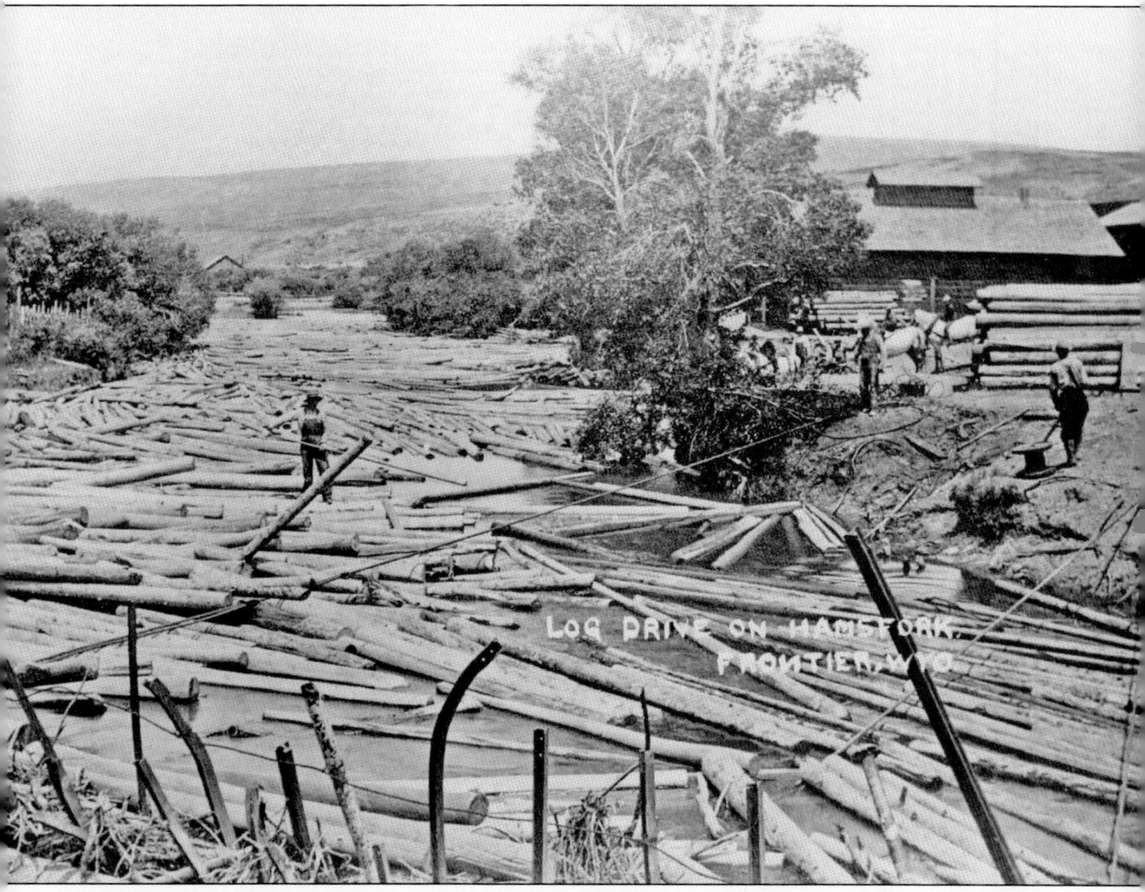

Log drives on the Hams Fork River started as early as 1880, when the Oregon Short Line Railroad began building tracks through the area. Timber was cut all winter, and the logs were stacked on the riverbank until they could be driven down the river. These log drives had to be made when the river was just right, as soon as the ice thawed and before the flooding began. When the river was flooding, the banks could not be seen and the logs could wash out of the river. Many times, the logs would jam and stop moving. The drivers would have to search for the jam and pry it loose with pike poles. This was very dangerous work. There was always a cleanup crew who would follow the river at the end of the drive, locating many of the stranded logs and placing them back into the river. Even with the cleanup crew, many logs were lost.

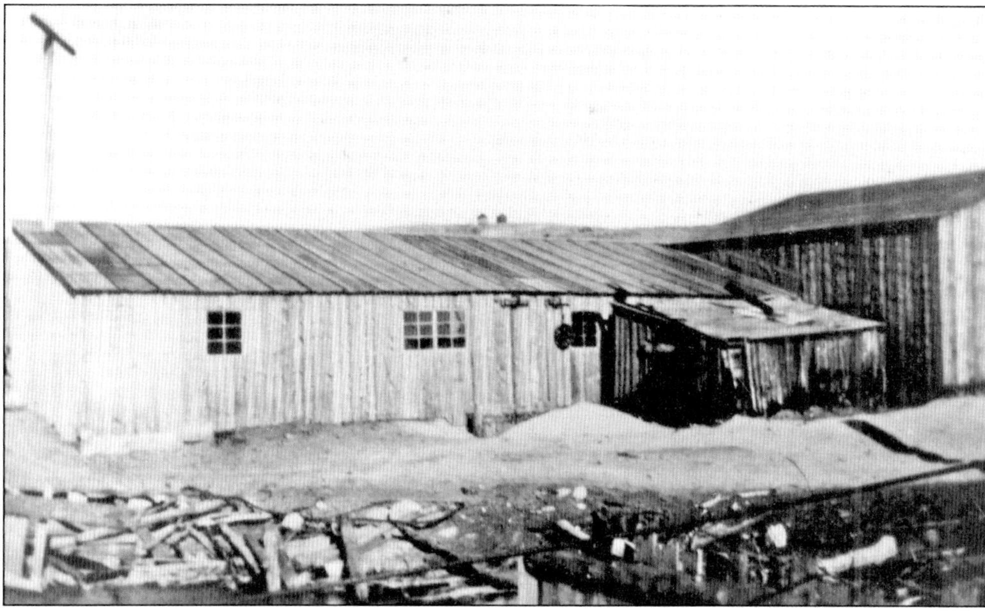

The Kemmerer Coal Company owned and operated the Frontier sawmill. The first log drive by the Kemmerer Coal Company was in 1908; the last was in 1921. Most of the timber was used for mine ties and props. A log boom was strung across the river, backing up the logs. Men with horses and grab hooks would pull the logs from the river and place them next to the sawmill. The men working at the sawmill were called log drivers and were employed by the Kemmerer Coal Company; they were able to live in Frontier in the company houses and pay the same rent as the miners.

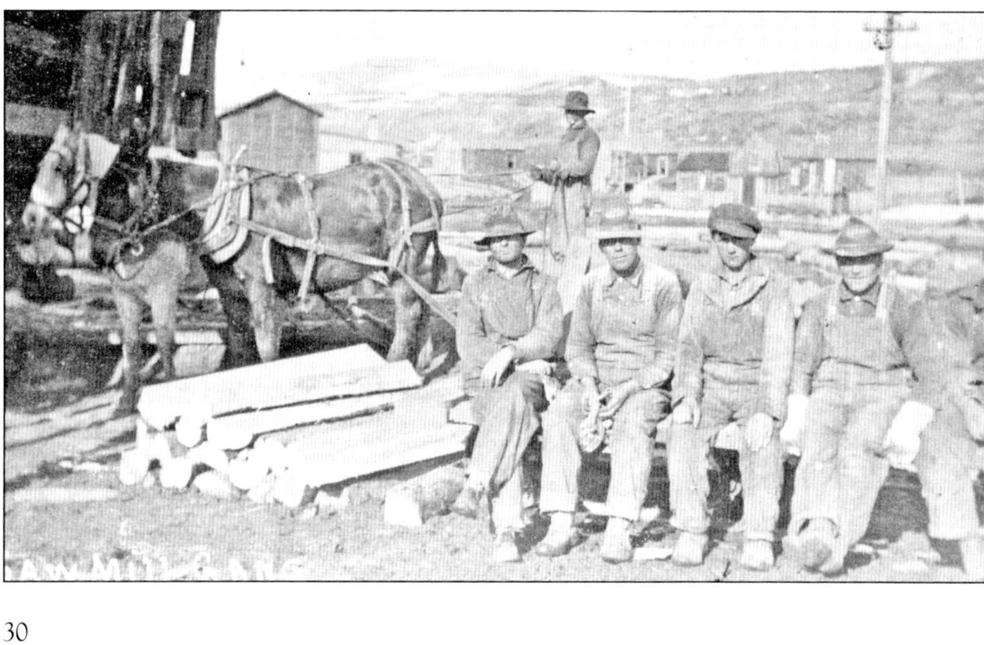

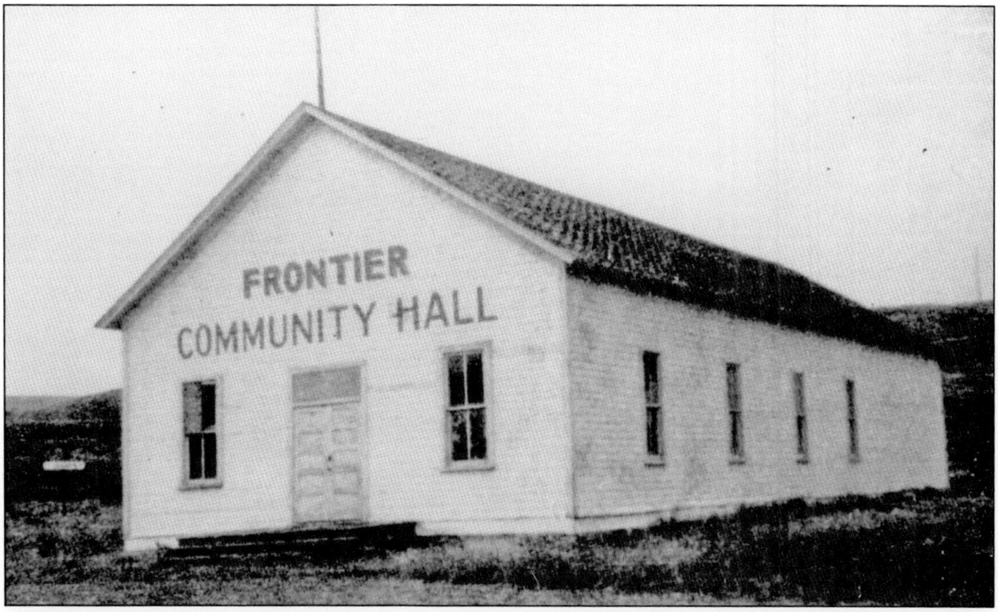

The Frontier Community Hall, or the Union Hall, was built in 1900 by the coal company. It was not used as a union hall until after 1903. The building was used for all forms of entertainment—plays and later movies, dances, weddings, and church services. More serious events, like town meetings and voting, were also held in this building. In 1918, the company remodeled the building, adding a heating system and indoor bathrooms. In 1925, the Frontier schoolhouse burned down, and the hall was used for school until a new one could be built. All other activities stopped taking place during this time. Dances and meetings were held in private homes. The popular Hood's Community Band would play for dances. If a man could play an instrument, he could join the band and be hired to work in the mines.

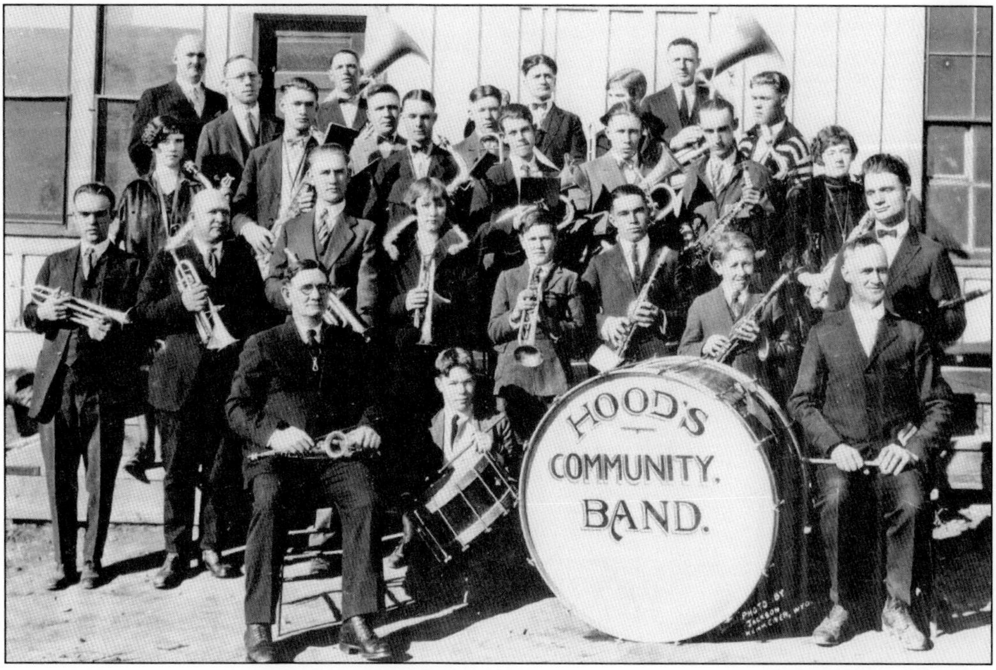

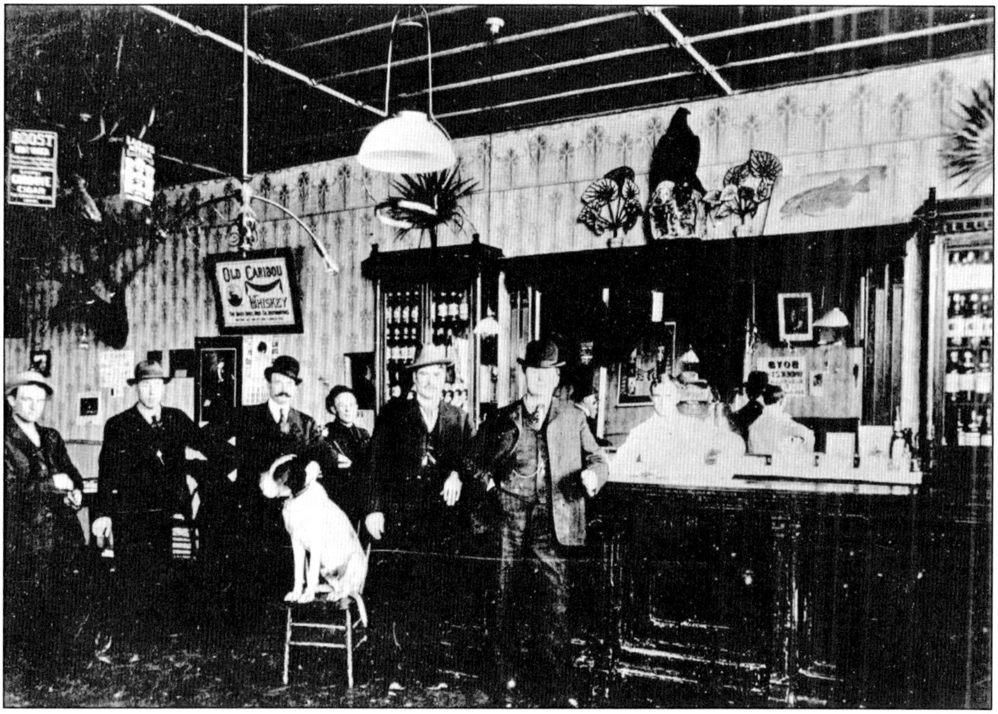

All the businesses in Frontier were owned and controlled by the Kemmerer Coal Company with the exception of the Frontier Saloon. It was moved to Frontier in 1898 from Twin Creek, a coal mining camp just west of Kemmerer. The building was a store and post office in Twin Creek run by Arthur Robinson. It was called the Alementation Saloon and carried this name for many years. The saloon has had many different names over the years since, including the Snake, Bucket of Blood, and finally the Frontier Bar and Restaurant. On April 12, 1899, Robinson sold the saloon, and by November 30, 1907, it was owned by Larry Quealy.

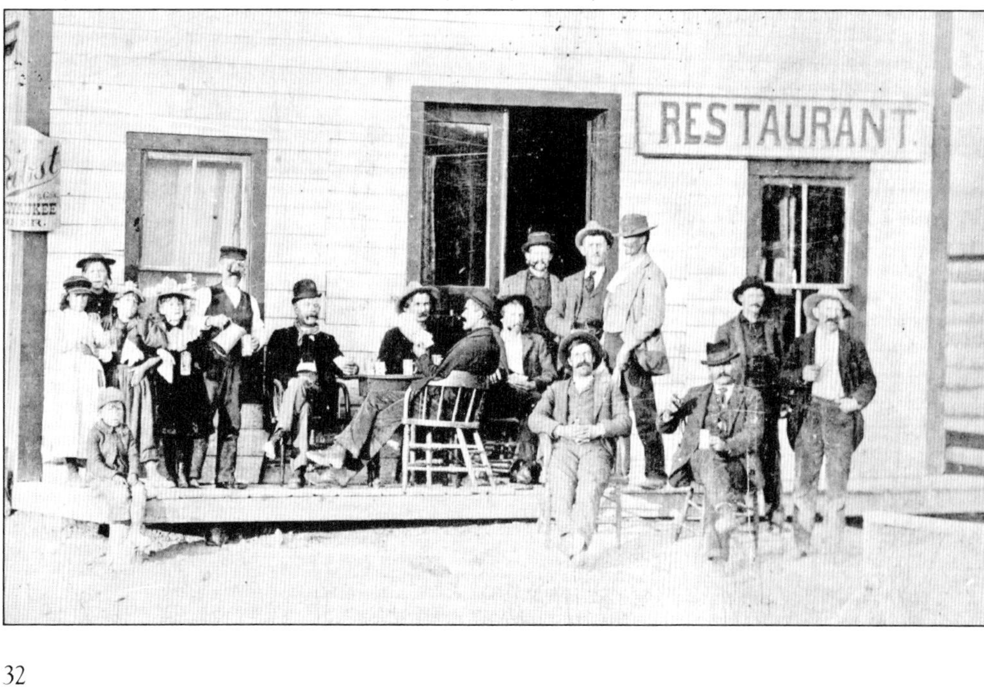

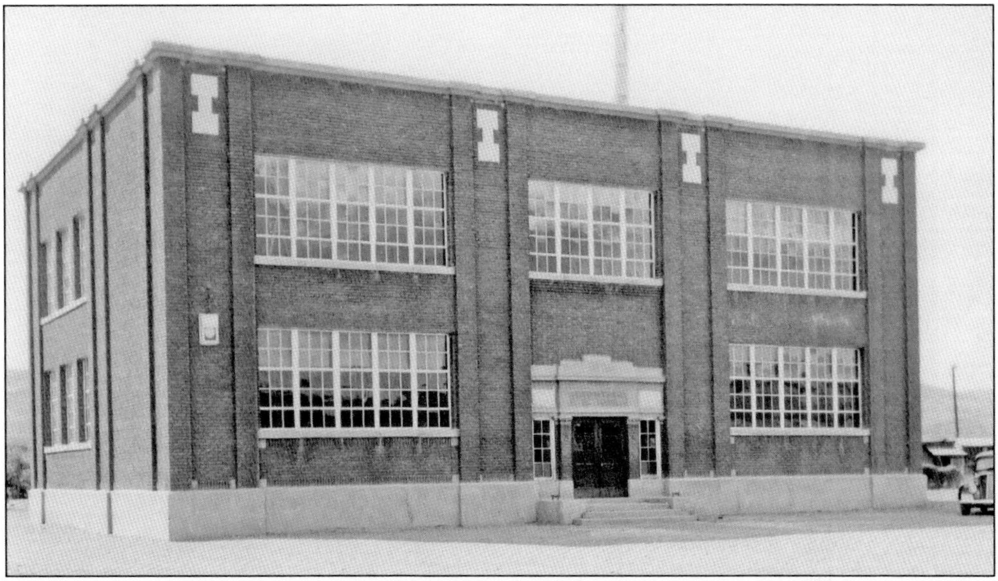

The first school in Frontier was built in 1900. It soon became too small and was replaced in 1901 by a larger wooden one. The building was placed on blocks, most likely so it would not be flooded by the Hams Fork River during the spring runoff. The boys would play marbles under the building. In 1924, the school burned down, and classes were held in the Frontier Community Hall and in Robinson Hall next to the Frontier Saloon until a new brick building was completed. Frontier students attended this school until 1965, when they were transferred to Kemmerer. During the early years, the classes were made up of mostly girls; very few young men attended because they were expected to help earn money for the family by working for the coal company. The young men pictured below are, from left to right, George Brown Jr., unidentified, Dubs Dubinsky, two unidentified, Ton Jiacoletti, Romeo Lupcho, and Eddie Cashin.

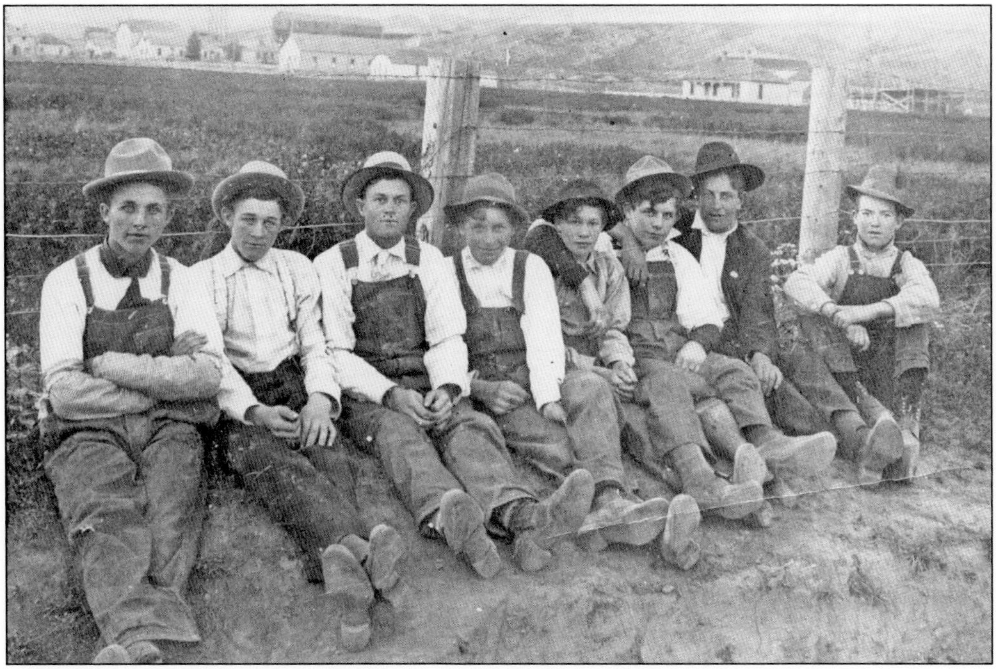

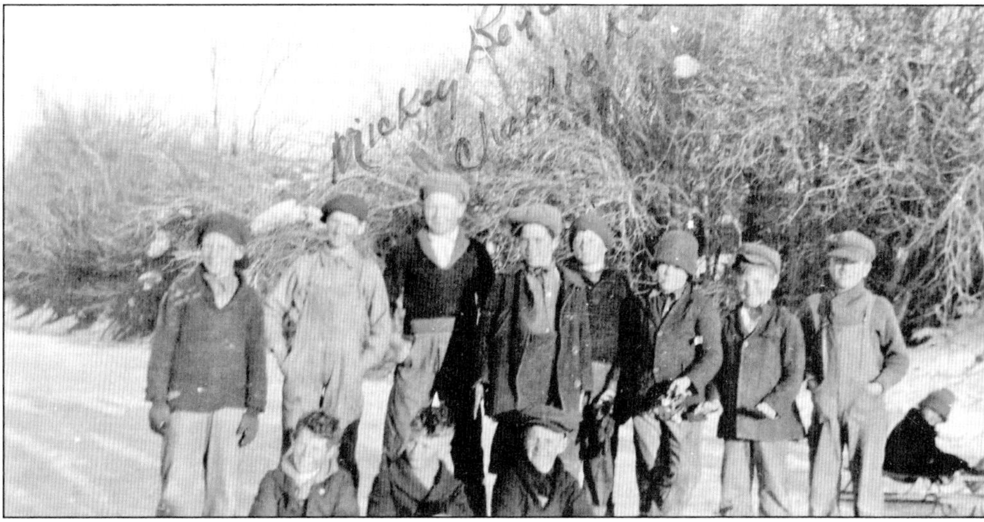

During the summer months, the families would gather for picnics, fishing, and swimming in the Hams Fork River. The occasional friendly baseball game, a favorite pastime, would follow in the evening. During the winter, the swimming hole would become an ice skating pond. The children would play ice hockey, ice skate, and sleigh ride.

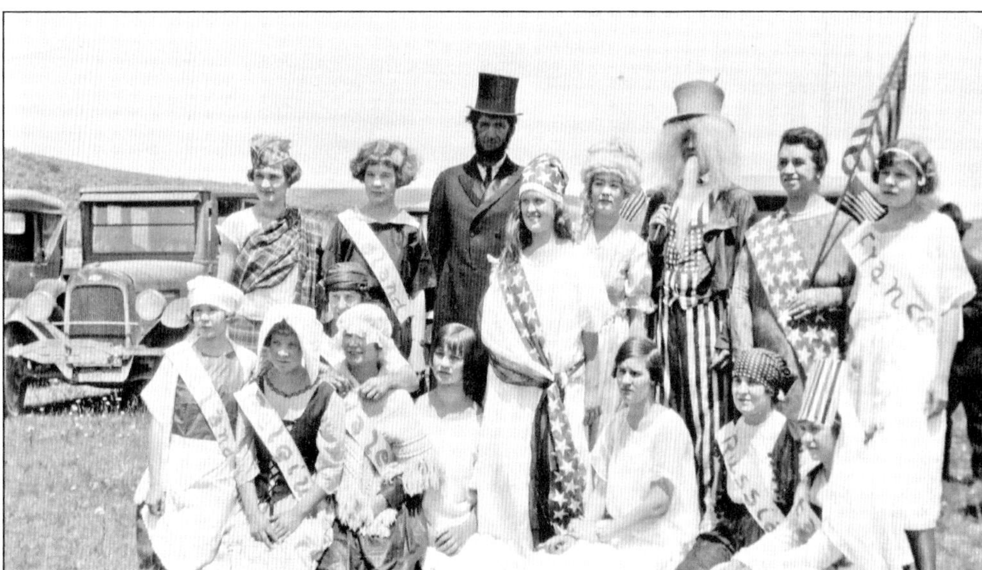

Fourth of July in Frontier was always a large celebration. The mines would be closed on that day so everyone could enjoy the day's festivities. Quealy would give a speech before the activities began. People would dress in costumes of their native countries and proudly wear a banner with the country's name on it. There would be a community picnic, contests, and a baseball game.

Three

MATURING COMMUNITY

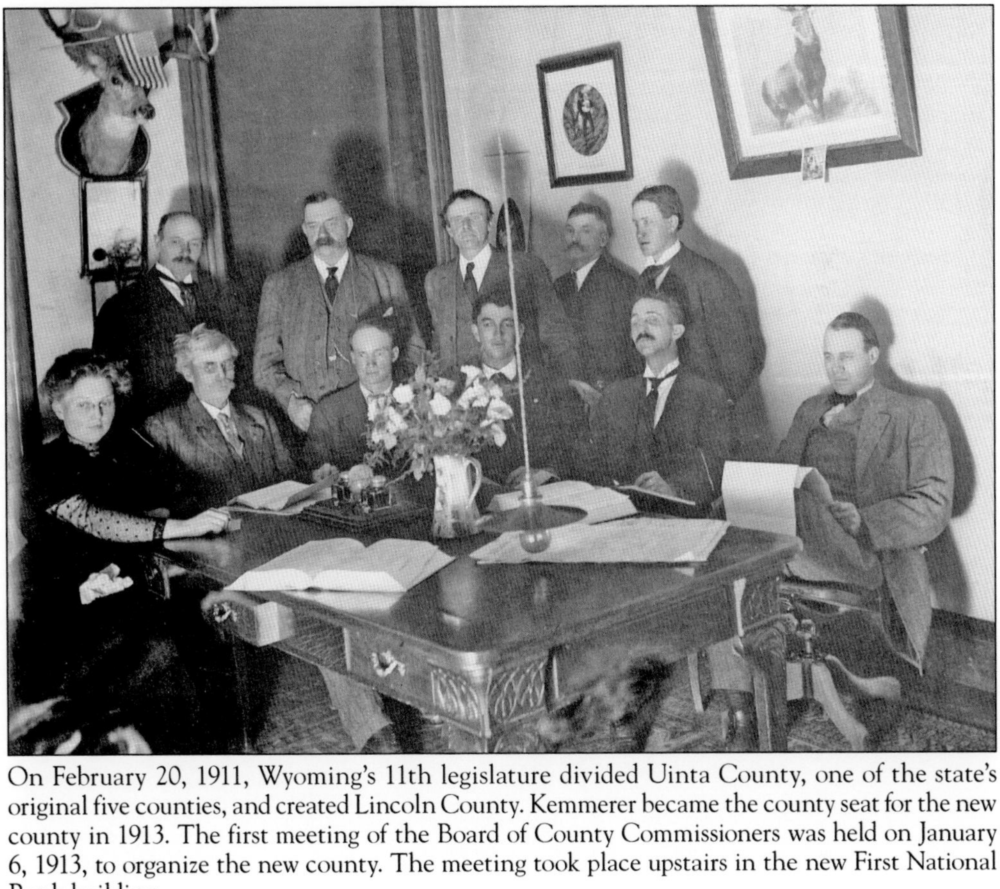

On February 20, 1911, Wyoming's 11th legislature divided Uinta County, one of the state's original five counties, and created Lincoln County. Kemmerer became the county seat for the new county in 1913. The first meeting of the Board of County Commissioners was held on January 6, 1913, to organize the new county. The meeting took place upstairs in the new First National Bank building.

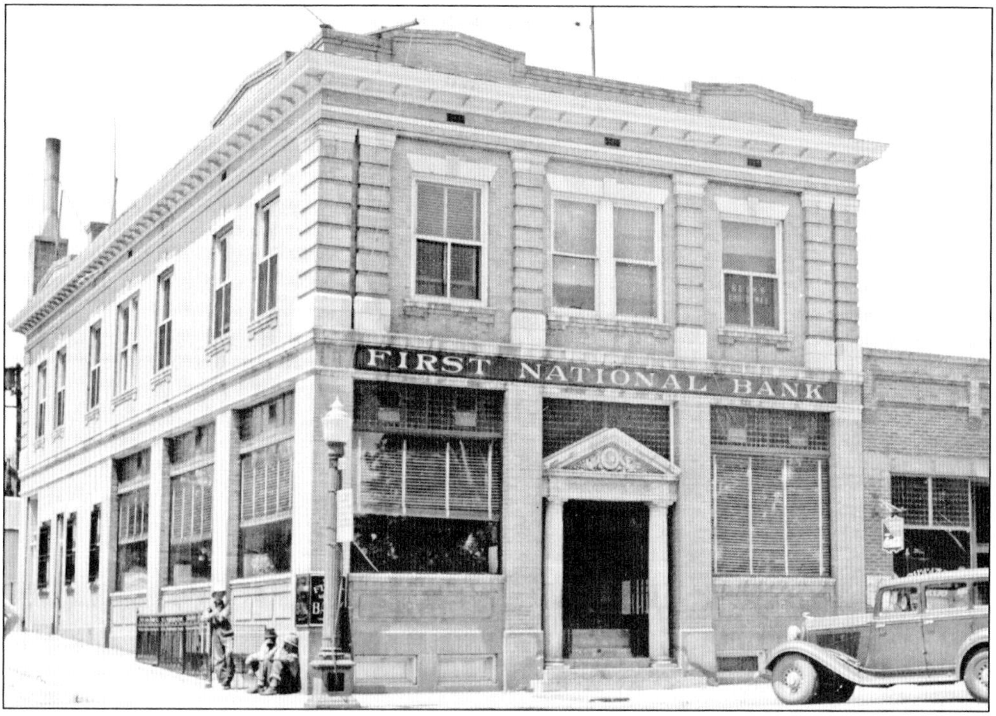

As the town of Kemmerer continued to grow, the bank also grew. The First National Bank had become one of the strongest financial institutions in the West. By January 1913, a new building had been built to house the new bank, down the street from the original bank building. The old building became the Kemmerer Savings and Loan and later the Ruth Elizabeth Shop. The new building housed the new county's offices until a courthouse could be constructed. There was also a barbershop in the basement with a side street entrance. The bank was remodeled in 1964.

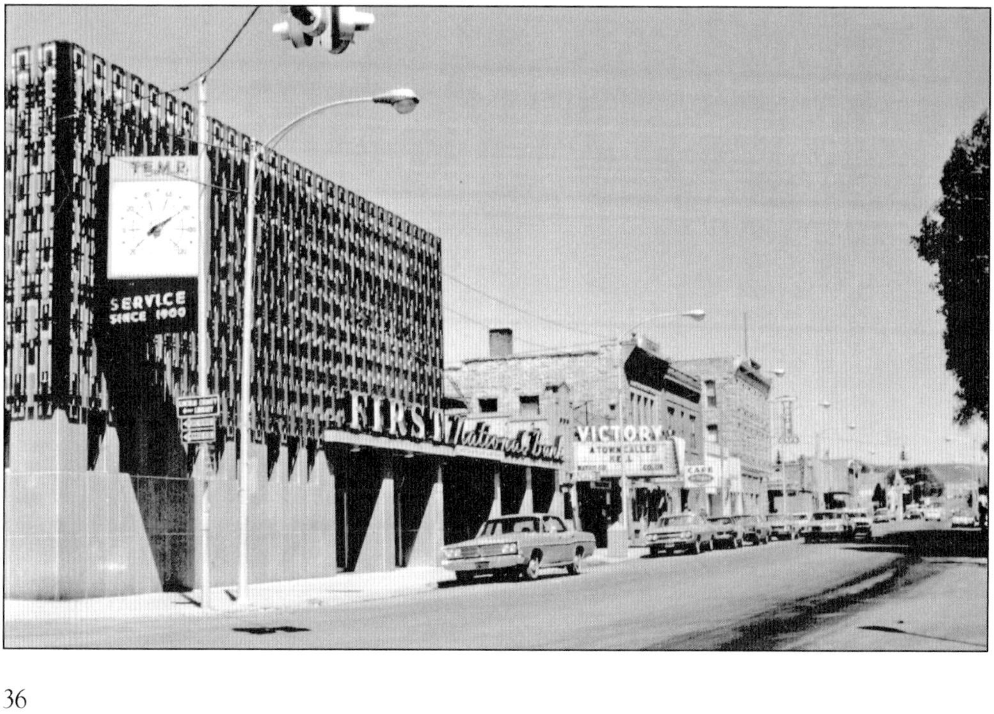

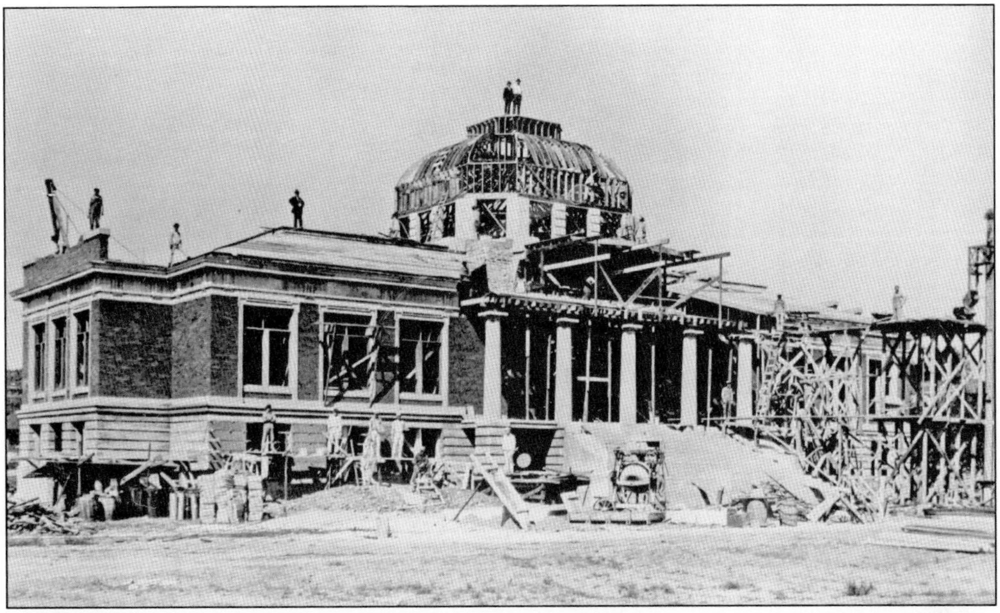

In 1924, Lincoln County Commissioners publicly expressed a need for a permanent building to house all the county offices, which were scattered throughout various Kemmerer businesses. There is no record of opposition to the courthouse project, and in August 1924, a bond issue in the amount of $100,000 easily passed. The construction got underway in the spring of 1925; the county offices were able to move into the new building just before Christmas of that year. It had served the needs of Lincoln County for almost 60 years with little change until 1981, when restoration work began to provide a growing county with a larger and more efficient building while still preserving the original integrity. The courthouse was built on land that had served as the first football field and also a baseball field.

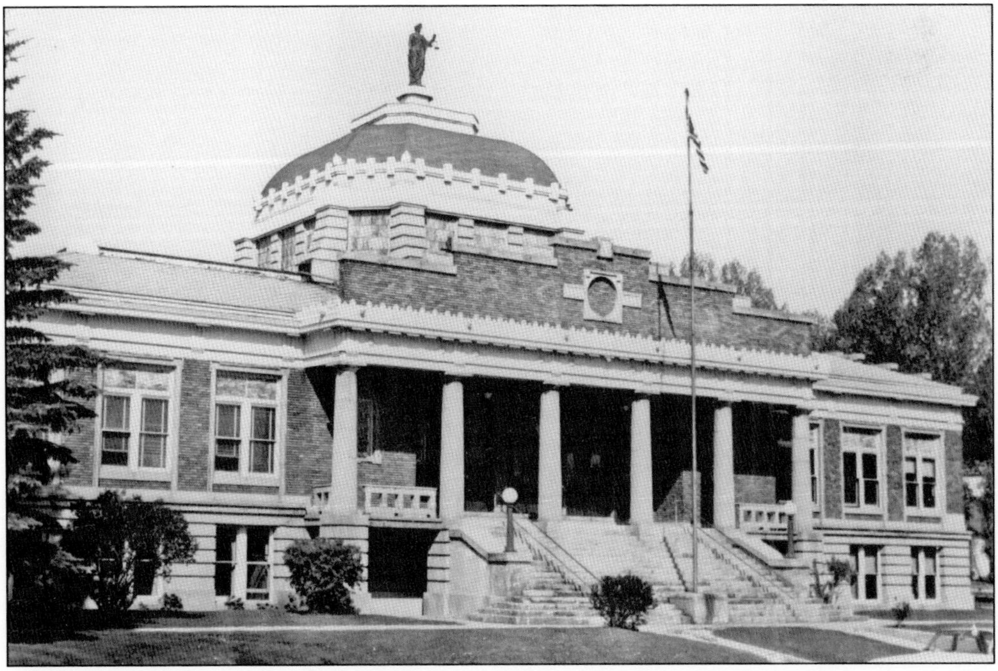

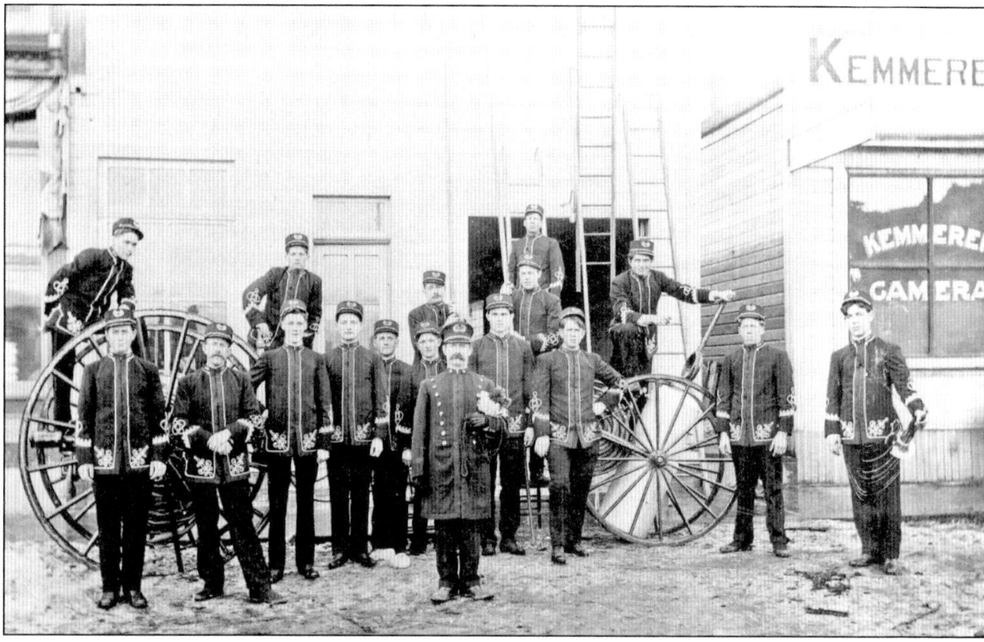

In August 1899, the Town of Kemmerer constructed a firehouse equipped with a hand-drawn fire "truck" (above). On October 6, 1902, Kemmerer had its first volunteer fire department, Kemmerer Hose Company No. 1. The company's uniforms were donated by Isabelle Burns, owner of the Southern Hotel. On May 12, 1916, the fire department purchased its first fire truck (below). In the fall of 1942, the American government was calling for scrap metal. The metal would be melted down and used to help in the war effort. The public was looking at sacrificing the old truck. Members of the fire department took the truck and hid it on a local ranch under a haystack. In 1970, the family gave the truck back to the fire department, and it is now on display at the fire station.

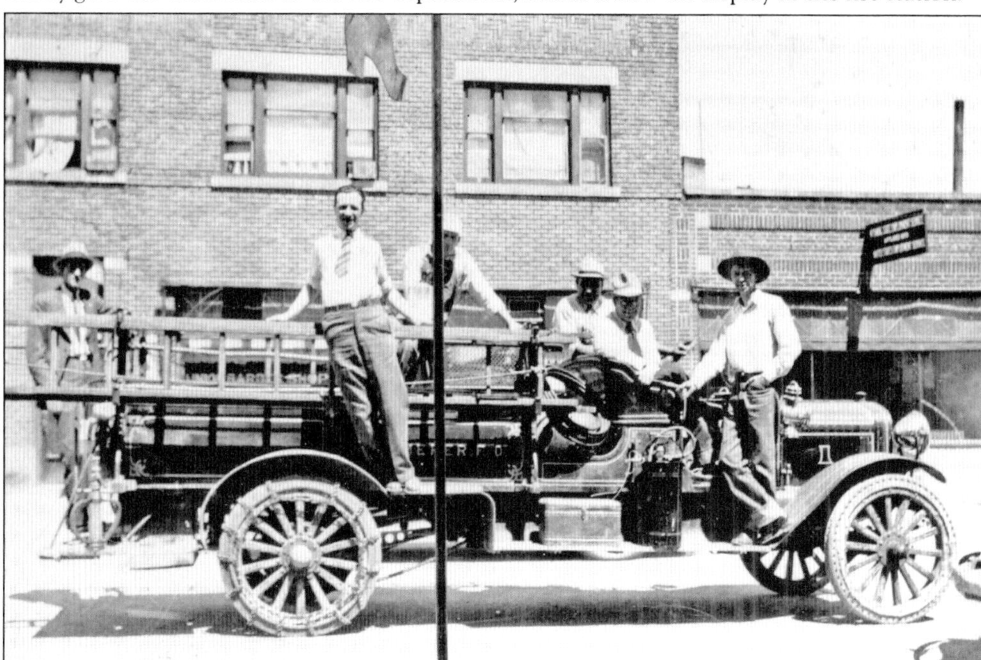

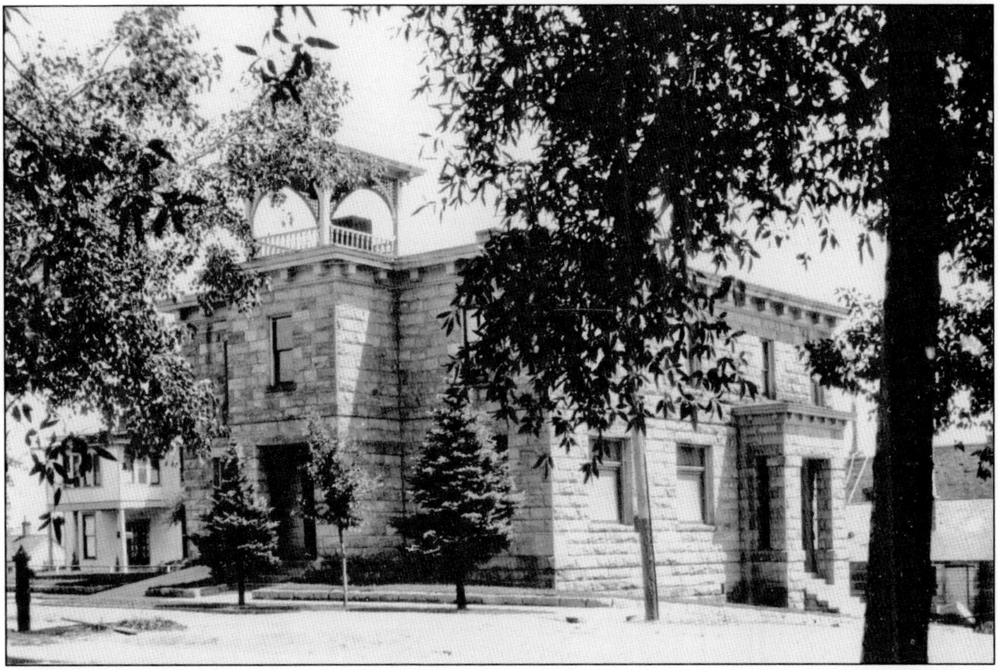

On February 28, 1899, the mayor and town council established a police department; the first town marshal was Mike F. Naylin. In August 1899, the new community discovered the need for a jail. The town's new jail (above) was completed on October 4, 1899. The county, which was still Uinta County, also built a jail with upstairs living quarters for the sheriff and his family. The county jail (below) was built in 1908. The town jail was enlarged in 1901 to make room for a city hall. The second story of the building held a courtroom. This is where the trial of cattle rustler Annie Richey was held.

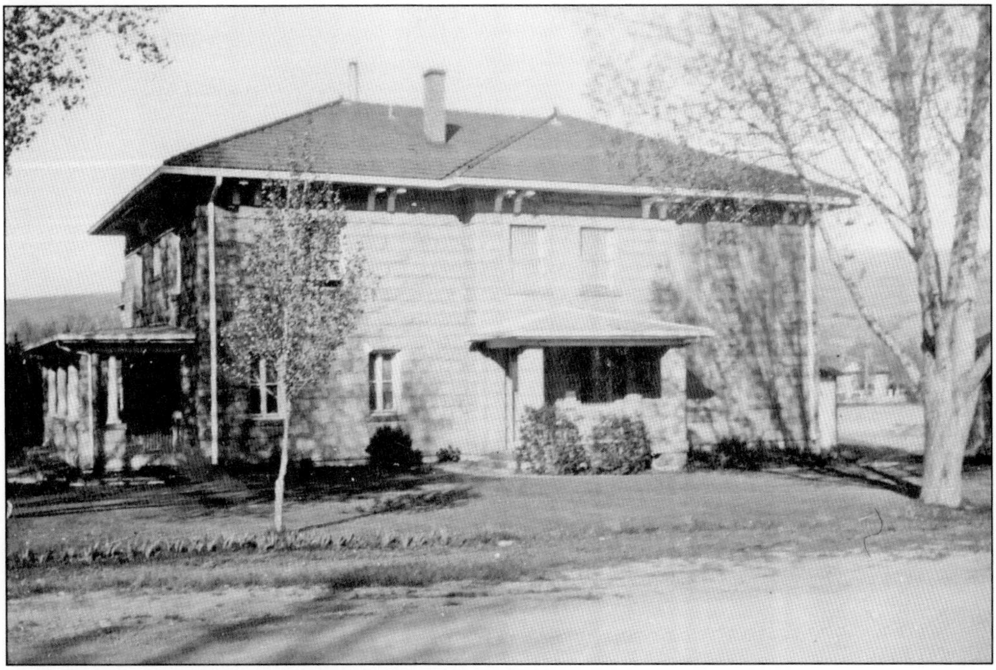

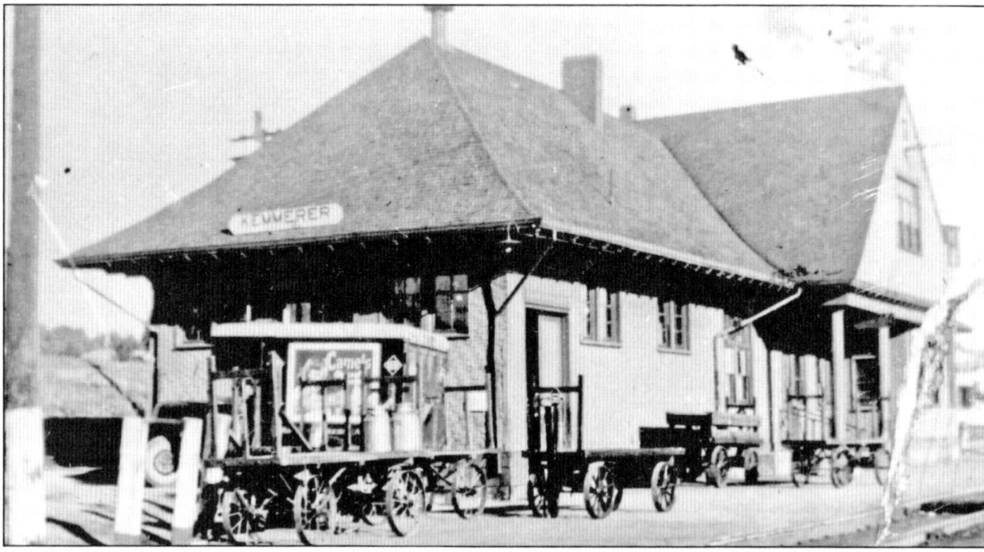

The Oregon Short Line Railroad was established on April 14, 1881. This railroad was a spur of the Union Pacific Railroad. The railroad was named the Oregon Short Line because it was built to provide the shortest route from Wyoming to Oregon. The building of the railroad was necessary because the Union Pacific's main line ended in Utah. Depots were a necessity for the railroad stops; Kemmerer's depot was built around 1898 on land that had belonged to the Kemmerer Coal Company. The depot was eventually torn down (the date is unknown). Kemmerer's place in railroad history is safe, and the opportunities for watching trains and train crews in action are still available today. Kemmerer's railroad depot stood for many years as a memorial to the many happy arrivals and sad farewells.

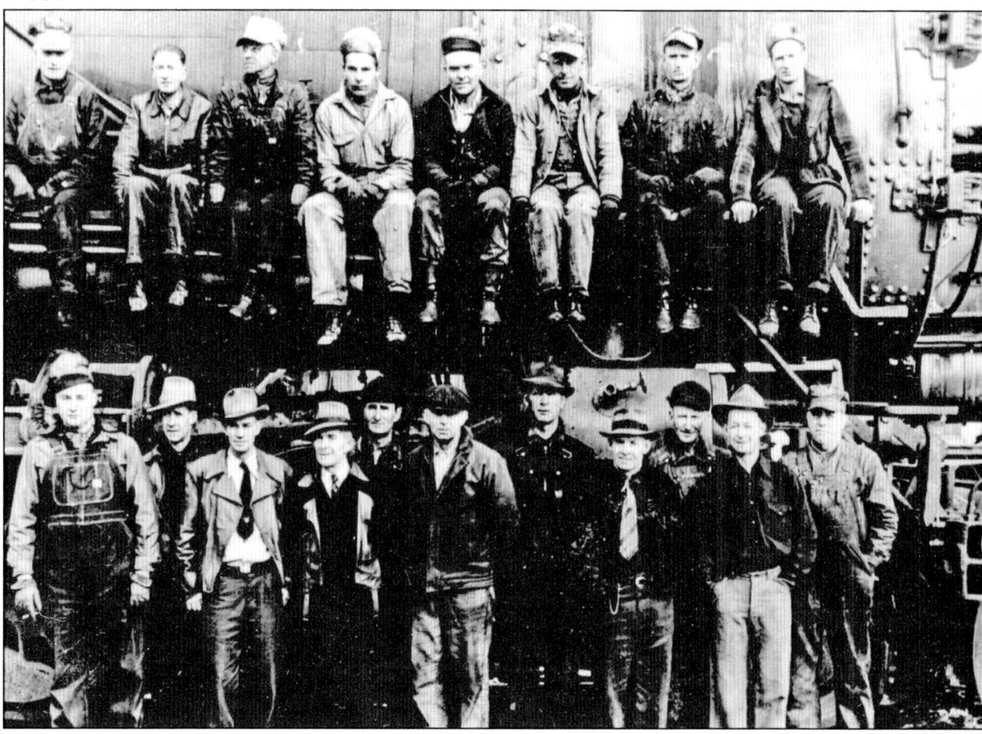

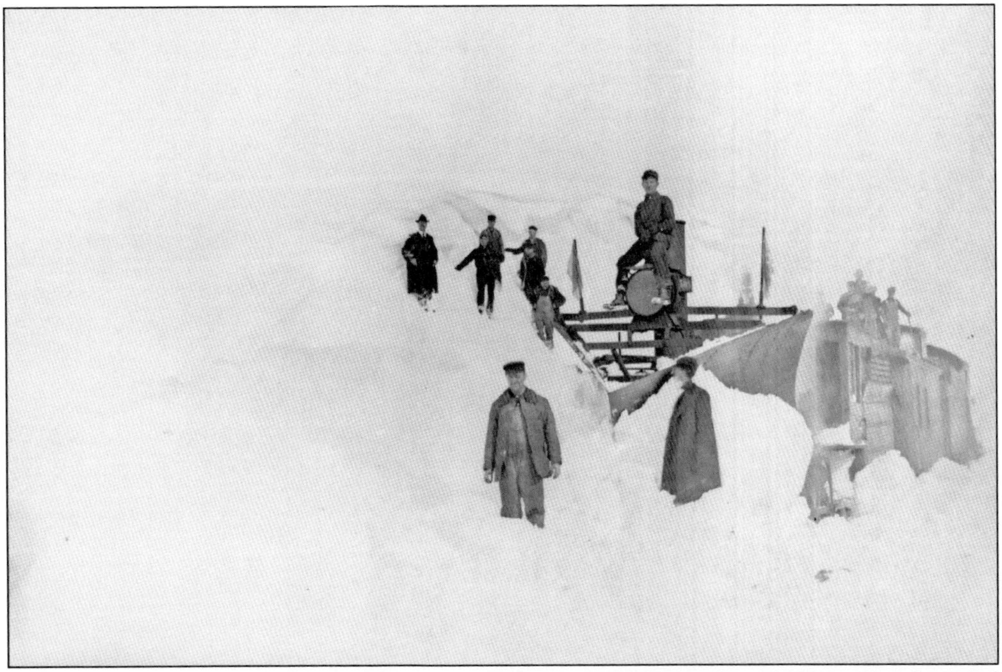

Like coal mining, the railroad played an important part in Kemmerer's past. The railroad needed the coal the mines produced, and the coal mines needed the railroad to haul the coal. The railroad also brought supplies to Kemmerer and the neighboring coal camps in the winter when the roads were impassable. During Prohibition, the trains would bring in large shipments of grapes and other supplies that were needed to make moonshine. The railroad crews would put large snow plows on the front of the steam engines to remove the snow from the tracks. Before the tunnel was dug at Tunnel Hill, large helper engines like these were used to push freight trains over the hill into Kemmerer.

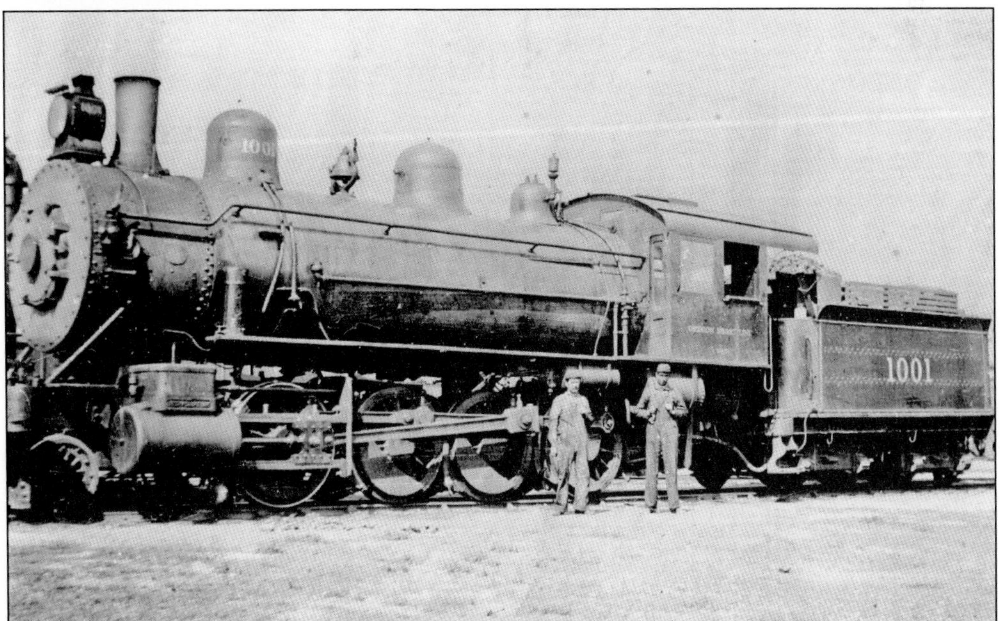

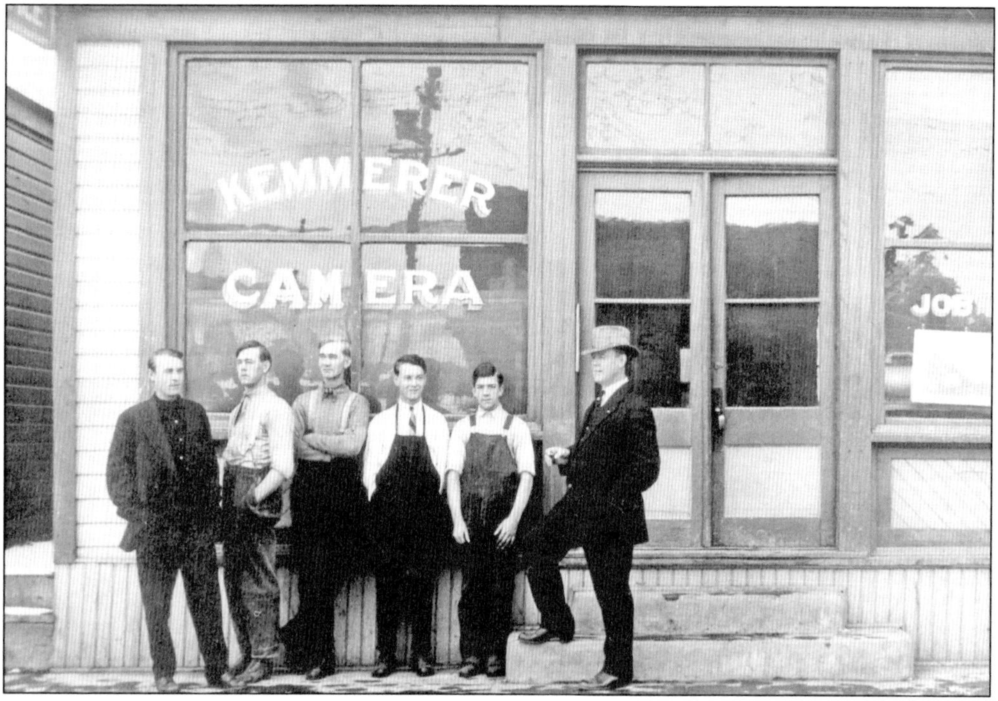

The Kemmerer Publishing Company was incorporated in 1901 and published the newspaper called the *Kemmerer Camera*. P. J. Quealy established the newspaper and used it to promote his point of view. By 1903, all of Quealy's other interests around Kemmerer were doing well except for the paper. There were 500 subscribers, but in 1908, financial and management problems forced the sale of the paper.

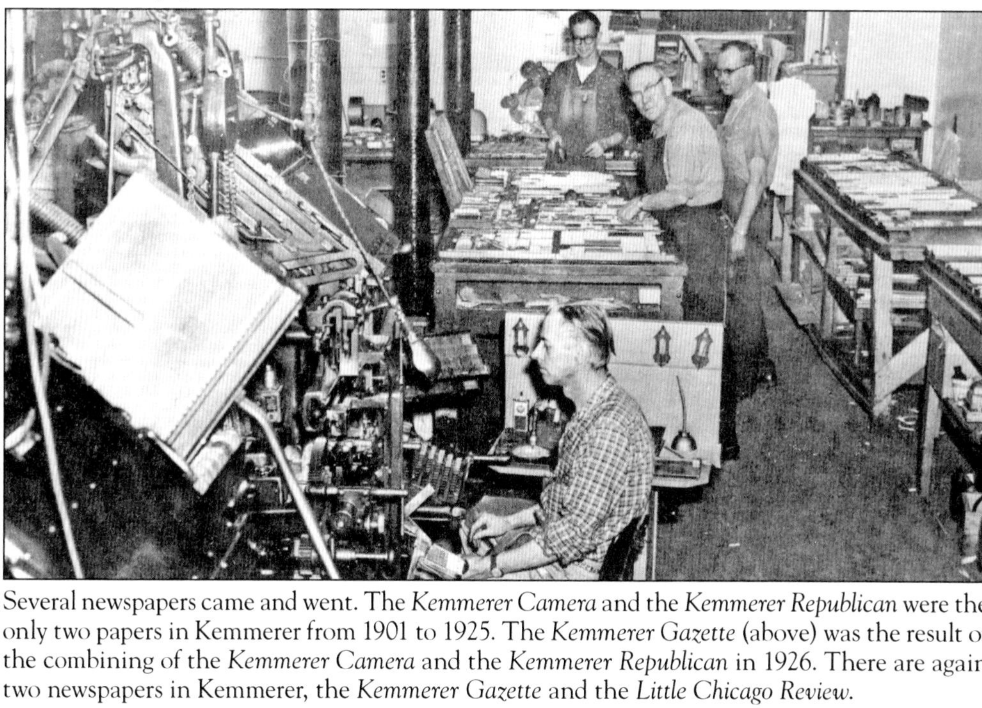

Several newspapers came and went. The *Kemmerer Camera* and the *Kemmerer Republican* were the only two papers in Kemmerer from 1901 to 1925. The *Kemmerer Gazette* (above) was the result of the combining of the *Kemmerer Camera* and the *Kemmerer Republican* in 1926. There are again two newspapers in Kemmerer, the *Kemmerer Gazette* and the *Little Chicago Review*.

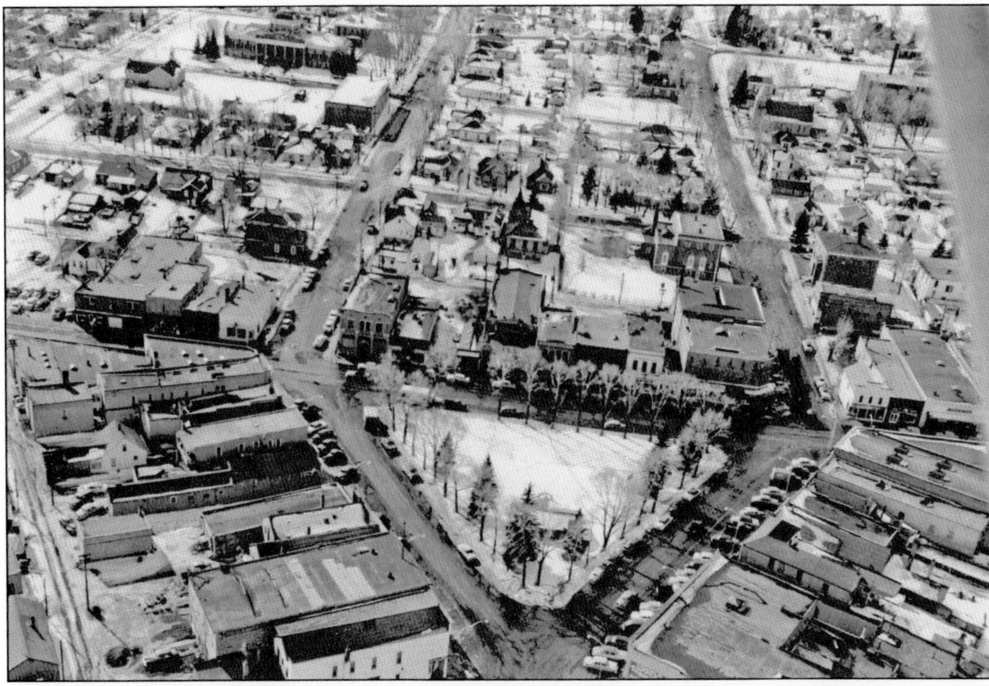

The Short Line Land and Improvement Company donated land to the town of Kemmerer in 1899 for a small park in the center of town. When the downtown area was laid out, it was discovered that the piece of ground was in the shape of a triangle. This 1950 photograph clearly shows the Triangle Park. Kemmerer is the only town in Wyoming whose town square is a triangle.

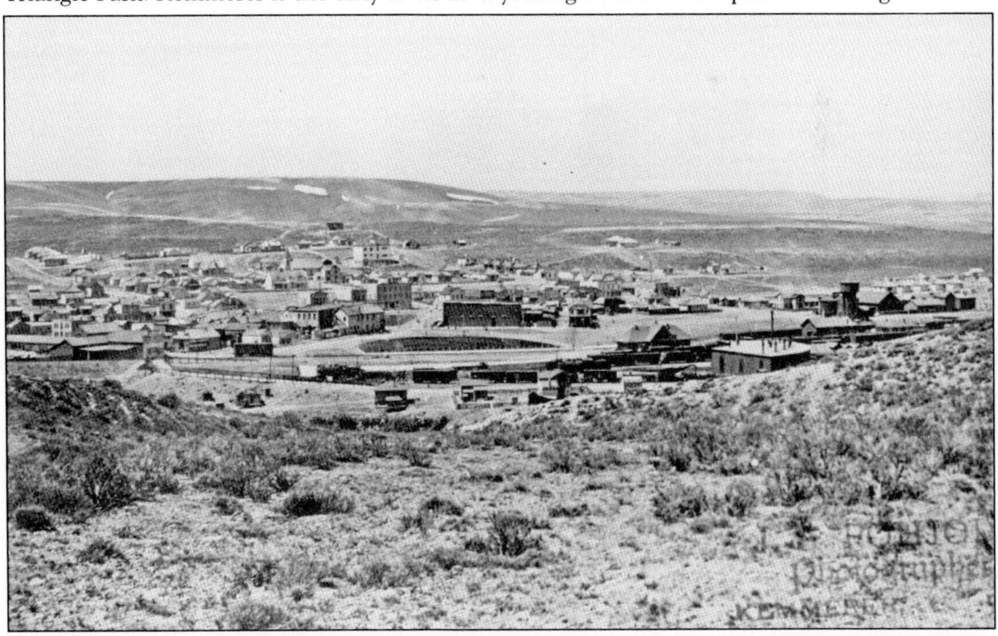

By 1902, Kemmerer was well on its way to becoming a thriving community. It was a business and railroad center for miners, ranchers, and business professionals. By 1915, Kemmerer had grown into a large community with several stores, hotels, a bank, churches, and schools. The values that shaped Kemmerer are still evident today.

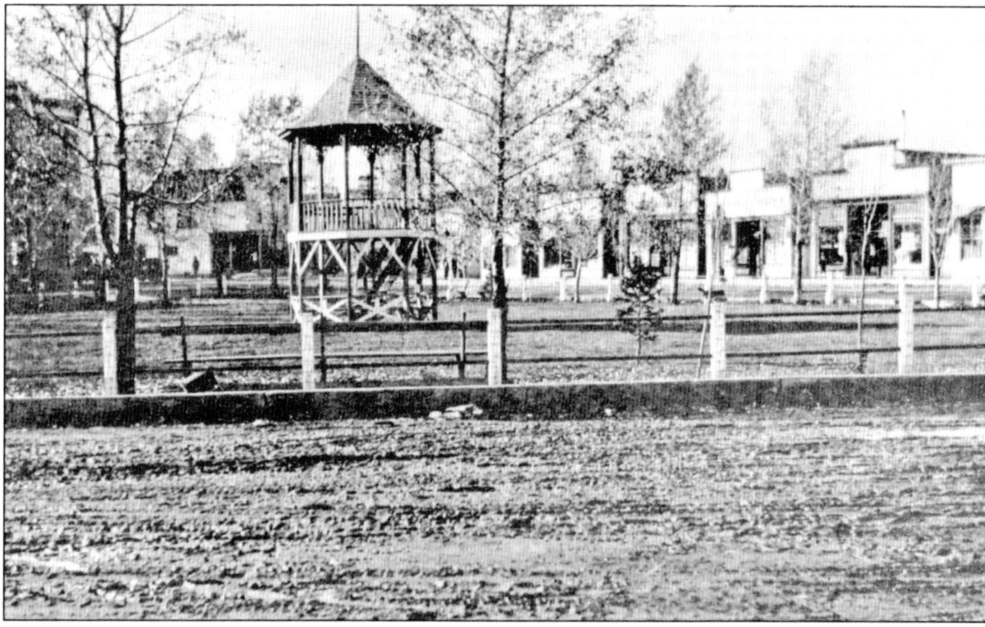

The Triangle Park from the beginning has always been a gathering place for community celebrations, meetings, and public mourning. In 1905, a gazebo and fence were built, making it more like a park. Before the town of Kemmerer was established, the area around the triangle had been a buffalo wallow. In 1899, the wallow was filled in and trees were planted. Board sidewalks were built so people could walk around the park without getting in the mud. Today the park continues to be a gathering place for the community, and in the summer, two large events are held there: Fossilfest and the Oyster Ridge Music Festival.

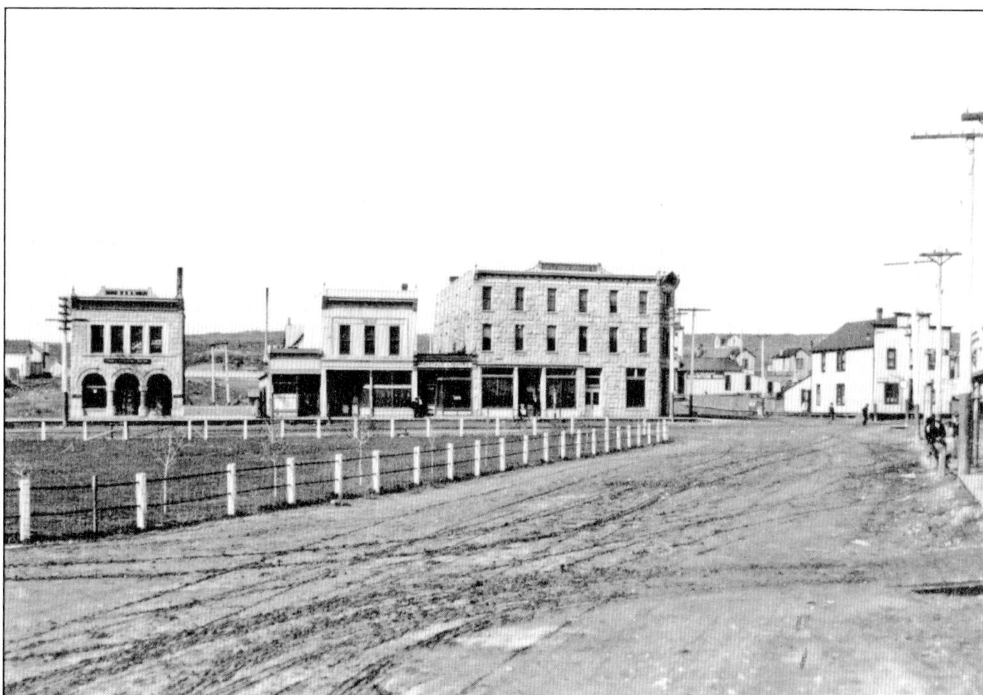

In 1924, this buffalo from Yellowstone National Park was donated to Kemmerer as meat for low-income families. After the head of the buffalo was prepared by a taxidermist, it had several homes in Kemmerer, including the First National Bank until 1959, when the bank was remodeled. The buffalo then went to Sawaya's Shoe Store. In 1989, Sawaya's donated the buffalo to the Fossil Country Museum, but it remains hanging in Sawaya's.

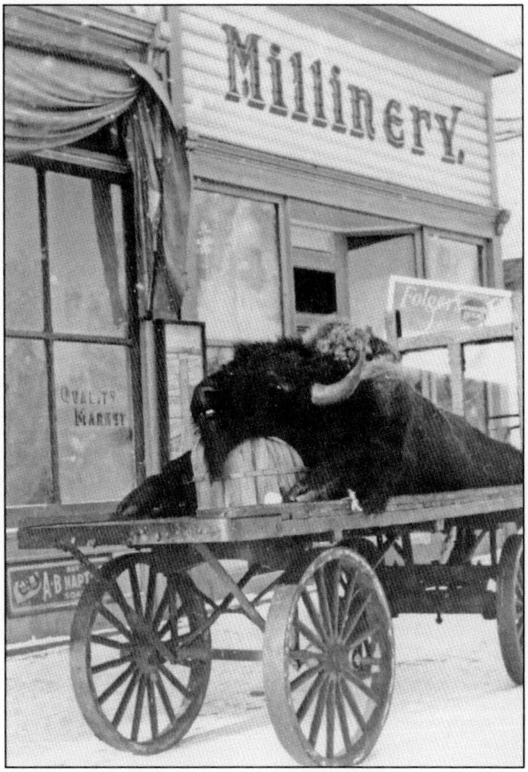

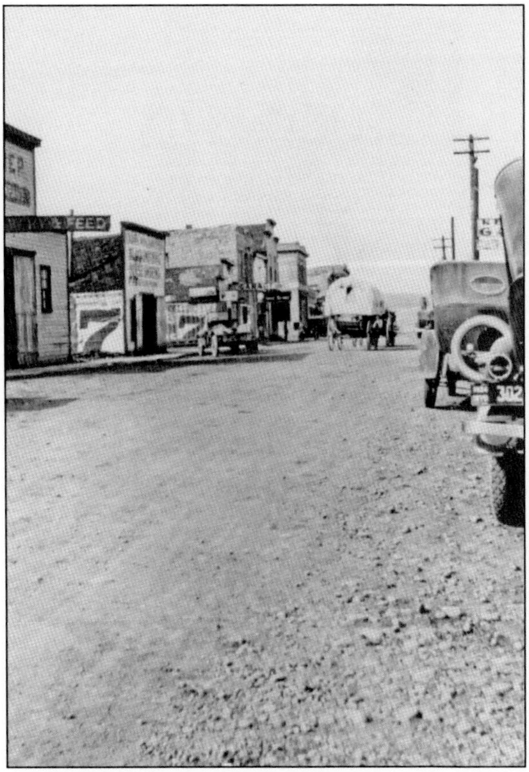

By 1924, automobiles had been introduced to Kemmerer, but the ranching families in the area still used wagons as their main means of transportation. The ranchers came to Kemmerer for their supplies; most trips would take two days, so they purchased enough supplies to last for a few months. The Frontier Supply Store would give the ranchers blankets with every purchase.

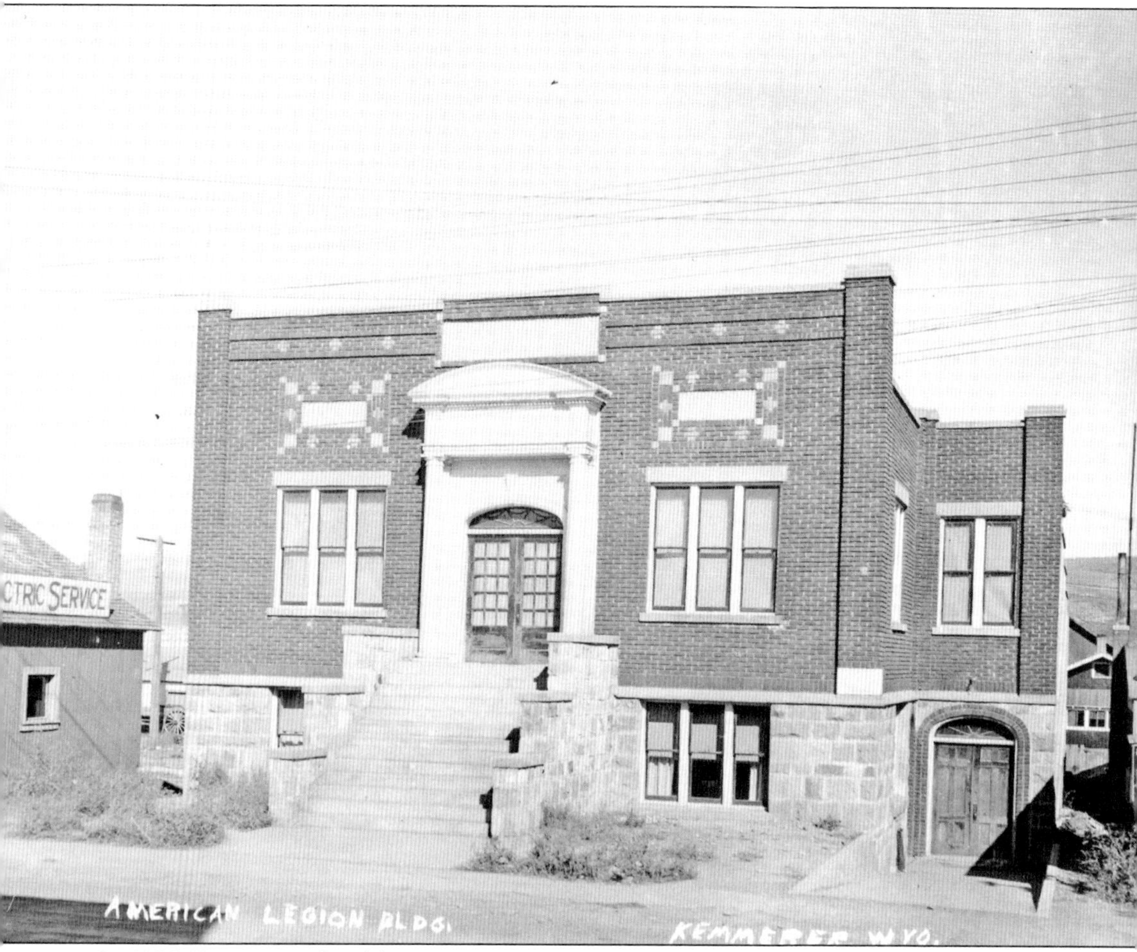

In 1926, Kemmerer's first library was constructed. J. C. Penney donated $10,000 for a memorial honoring his wife, Berta. Quealy donated a lot, and the public donated $15,000. It was decided to construct a building suitable for the library and an American Legion meeting place. The building was named the Berta Penney Memorial. The name of the building was eventually changed to the American Legion Hall after receiving a license to serve liquor. J. C. Penney did not want the Penney name associated with any building that served liquor. The rock foundation was donated from the old grade school building by the school district. In the 1920s, several people expressed an interest in a reading room, and from this group grew the Lincoln County Library Association. In 1926, the association wrote a constitution, bylaws, and articles of incorporation. At its first meeting, the new association was faced with finding a new place for the library, as the American Legion Hall was only temporary. In October 1926, the library moved into the basement of the new county courthouse.

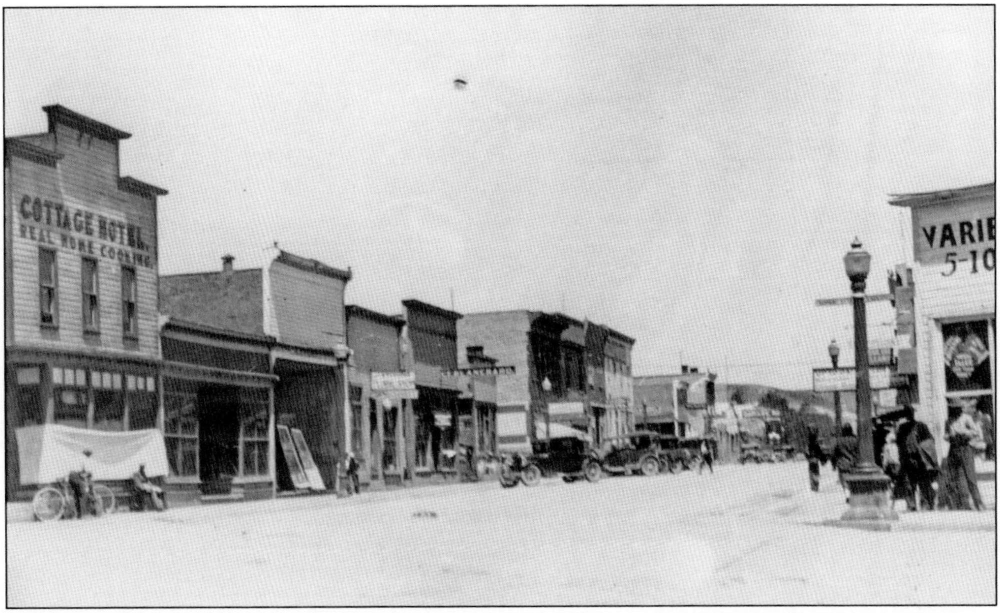

When word got out that a railroad station and a new town were beginning, hotels, most of them offering meals, were being built. Along with the Kemmerer Hotel, the Elk Hotel and Cottage Hotel, or Downey House, were some of the first to appear. A second Cottage Hotel, also known as the Hoskins Hotel or the Commercial Hotel, was located next to the Oregon Short Line Railroad Depot. This was one of the buildings moved to Kemmerer from the old town of Hams Fork. Kemmerer and Frontier both had rooming houses for the single miners, and most of the saloons in Kemmerer had rooms to rent.

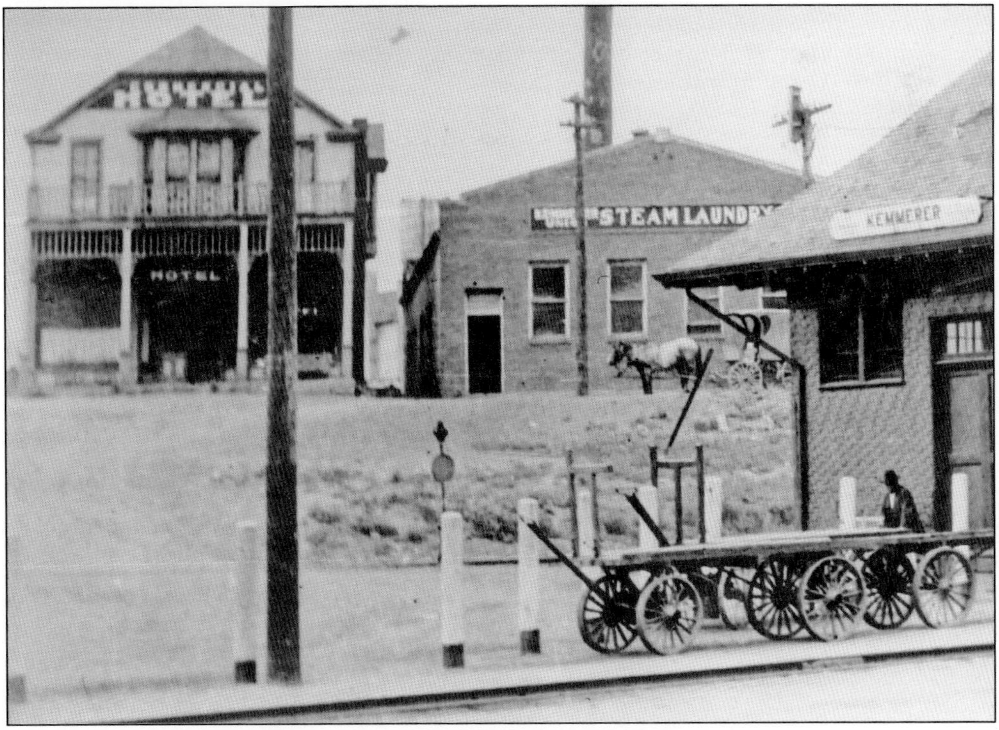

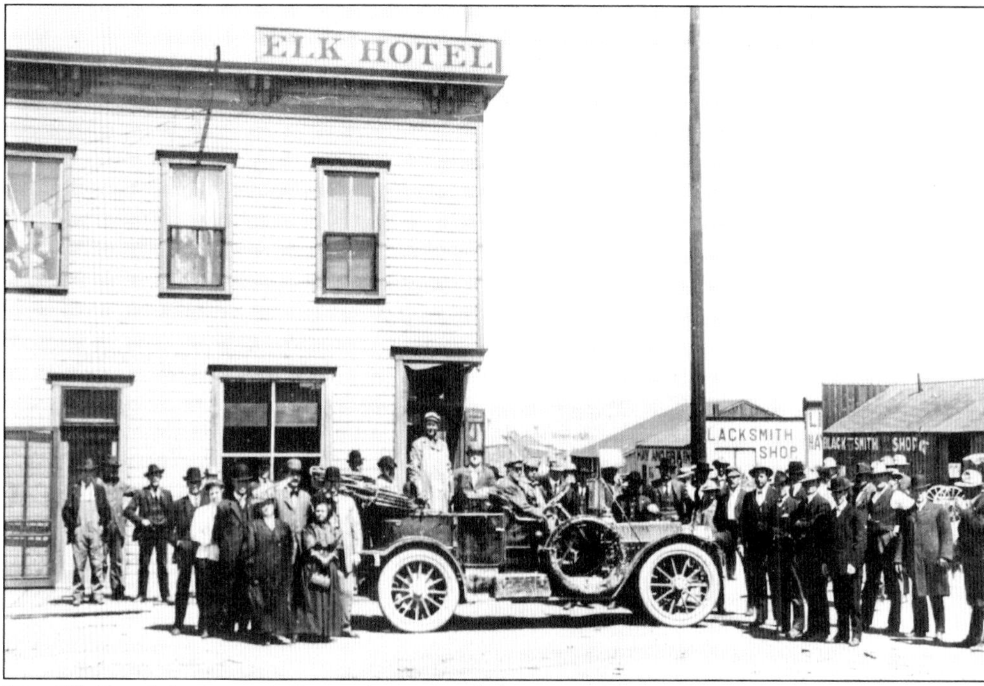

The Elk Hotel, built in 1897, was first named the Emerald Hotel. During the early days of its existence, the hotel catered to railroad men and ranchers. The hotel dining room always provided a hot meal for the working men. In 1902, the first automobile to come through Kemmerer stopped overnight at the Elk Hotel. This hotel was also where the jury was housed during the trial of cattle rustler Annie Richey. The French Hotel is in the original location of the first J. C. Penney store on Pine Avenue. The dates when the buildings were torn down are unknown.

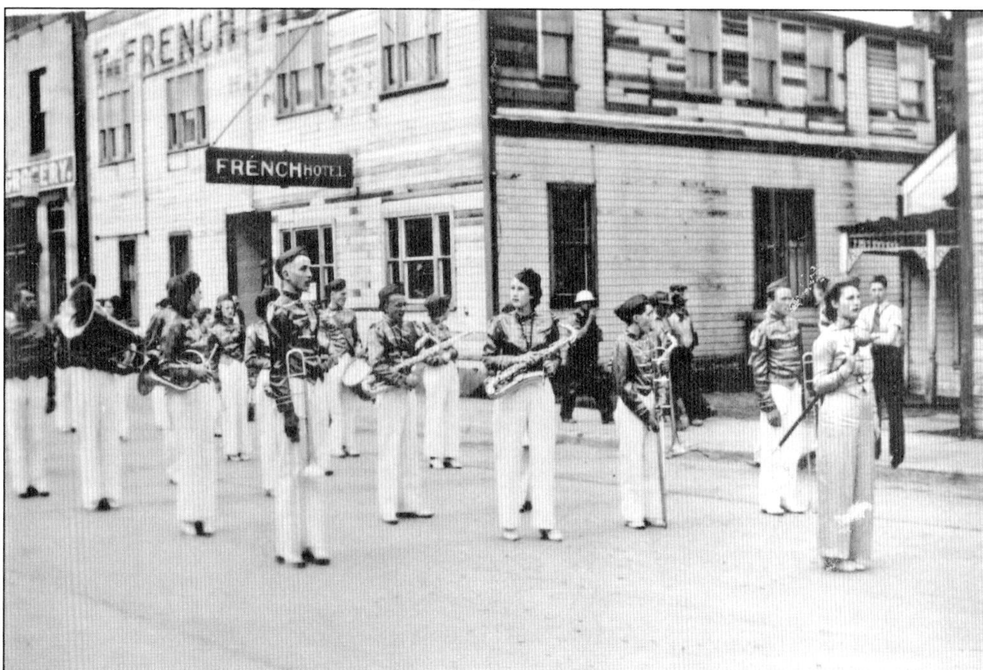

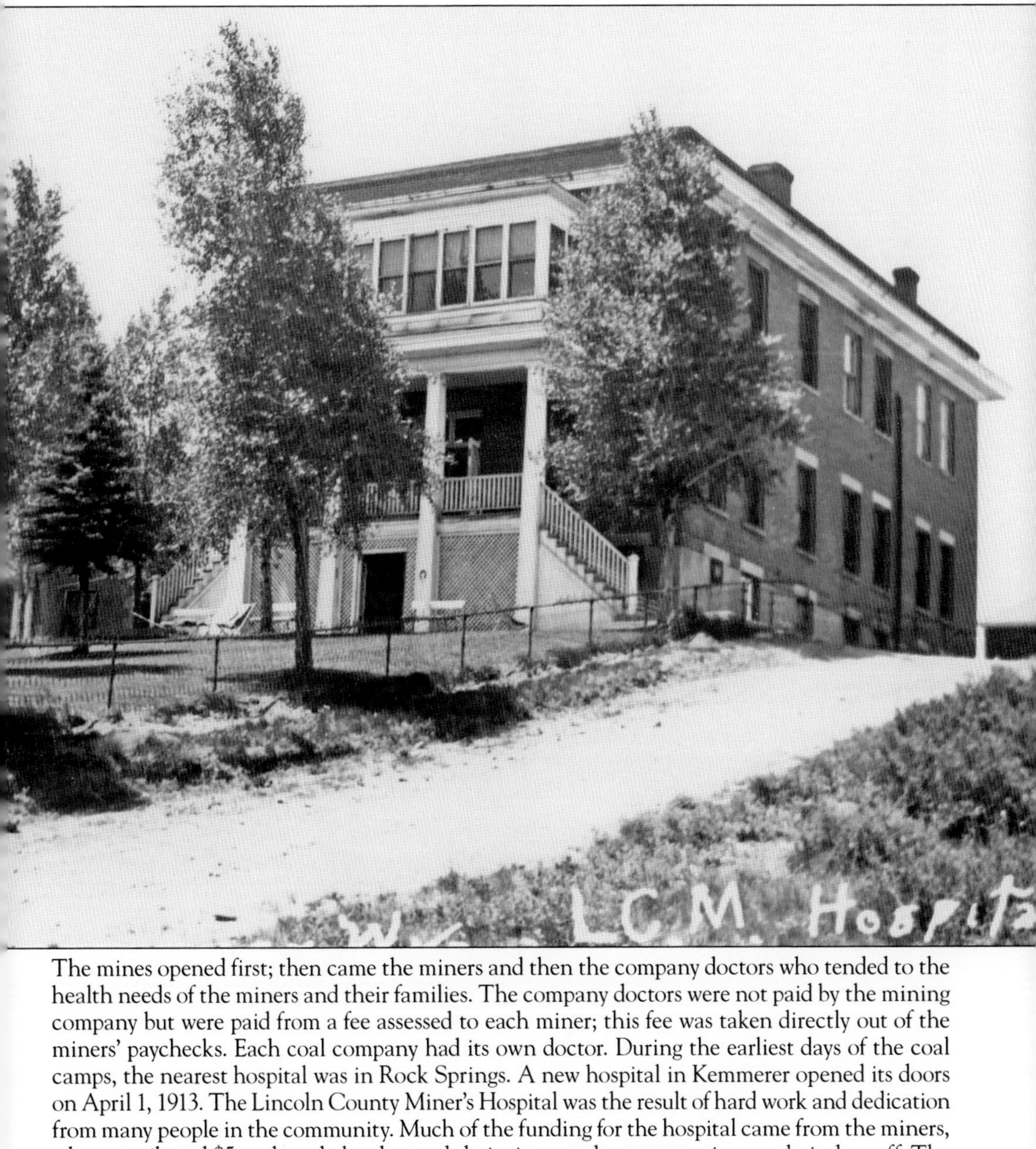

The mines opened first; then came the miners and then the company doctors who tended to the health needs of the miners and their families. The company doctors were not paid by the mining company but were paid from a fee assessed to each miner; this fee was taken directly out of the miners' paychecks. Each coal company had its own doctor. During the earliest days of the coal camps, the nearest hospital was in Rock Springs. A new hospital in Kemmerer opened its doors on April 1, 1913. The Lincoln County Miner's Hospital was the result of hard work and dedication from many people in the community. Much of the funding for the hospital came from the miners, who contributed $5 each and also donated their time on the construction on their days off. The hospital was governed by representatives from each of the nine United Mine Workers' locals. On August 18, 1959, a decision was made to transfer operations and equipment of the miners' hospital to a taxing entity, and the South Lincoln Hospital District was formed.

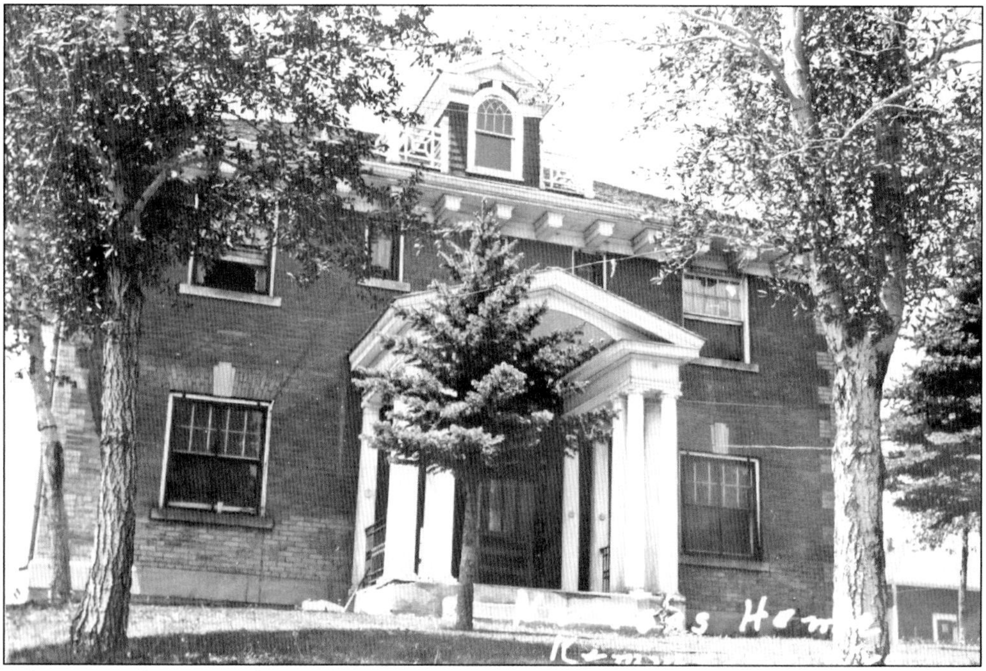

Adjacent to the miners' hospital was a nurses' training school and living quarters that operated until 1920. It was built in 1912 when the hospital was constructed. Gertrude Kemmerer Thorne, daughter of Kemmerer cofounder Mahlon Kemmerer, provided the funding for the building and furnishings for the home as a memorial to her mother. When the young nurses worked at the miners' hospital while in its four-year training program, they began with a salary of $10 per month; each year after, their salary increased by $1 per month. In 1910, Dr. John Marquis came to Kemmerer and opened a small office. The office was later made into a hospital, called the Marquis Hospital. Dr. Marquis also served three terms as mayor of Kemmerer.

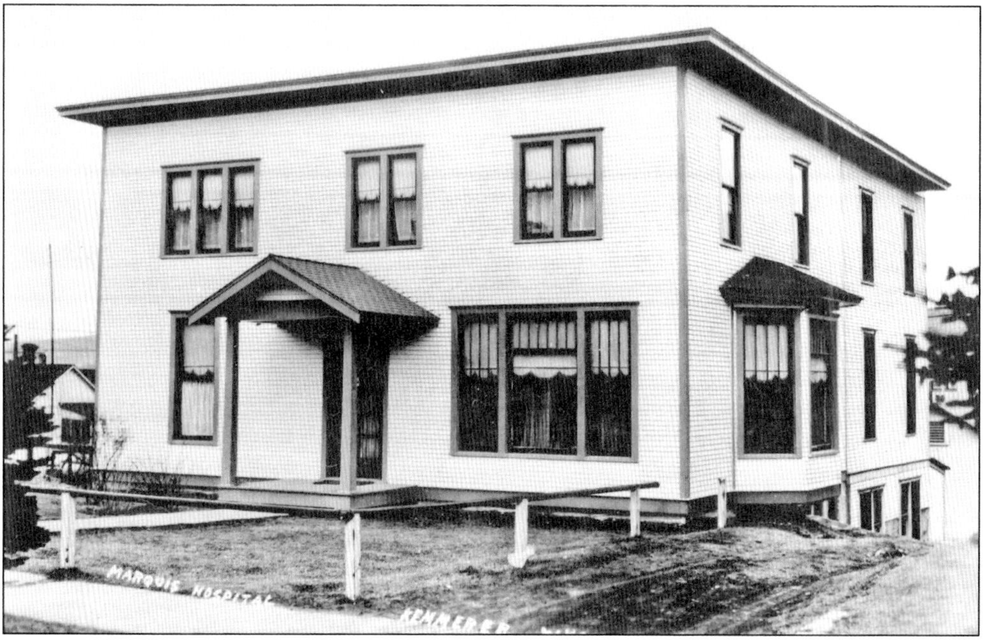

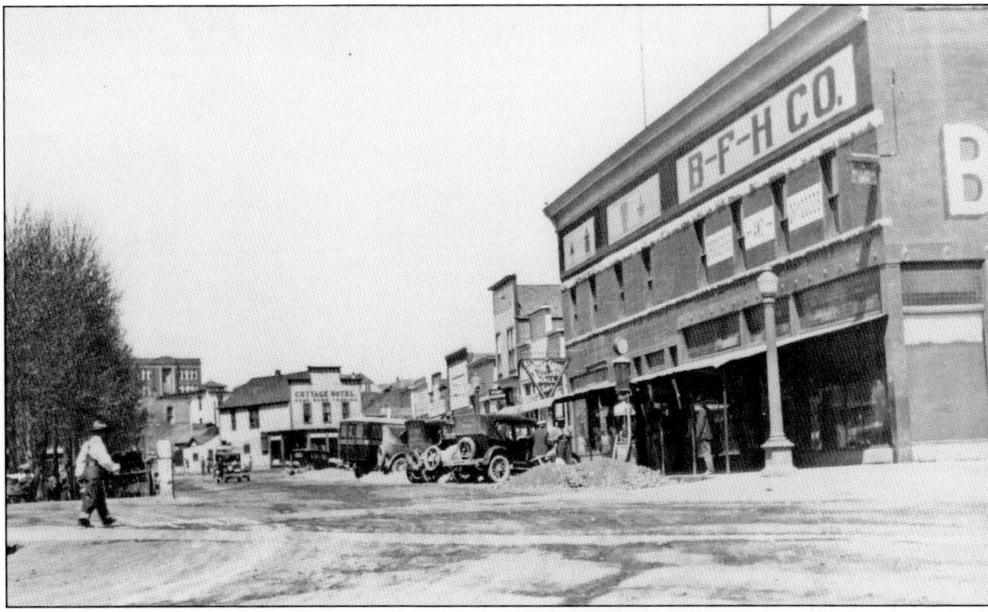

Located on north Main Street, now J. C. Penney Drive, was the Blythe-Fargo-Hoskins Company store. The first store to be built in Kemmerer, it was owned by William S. Post, Kemmerer's first postmaster. The store was one of the largest general merchandise establishments in the area. Almost anything a customer wanted could be found in the store. It continued to grow and eventually had several branches in the neighboring communities. This store was popular with the ranching families who would come to Kemmerer to shop because it had a place to keep wagons and horses while people shopped. The building held many different businesses over the years; it burned to the ground in the 1990s.

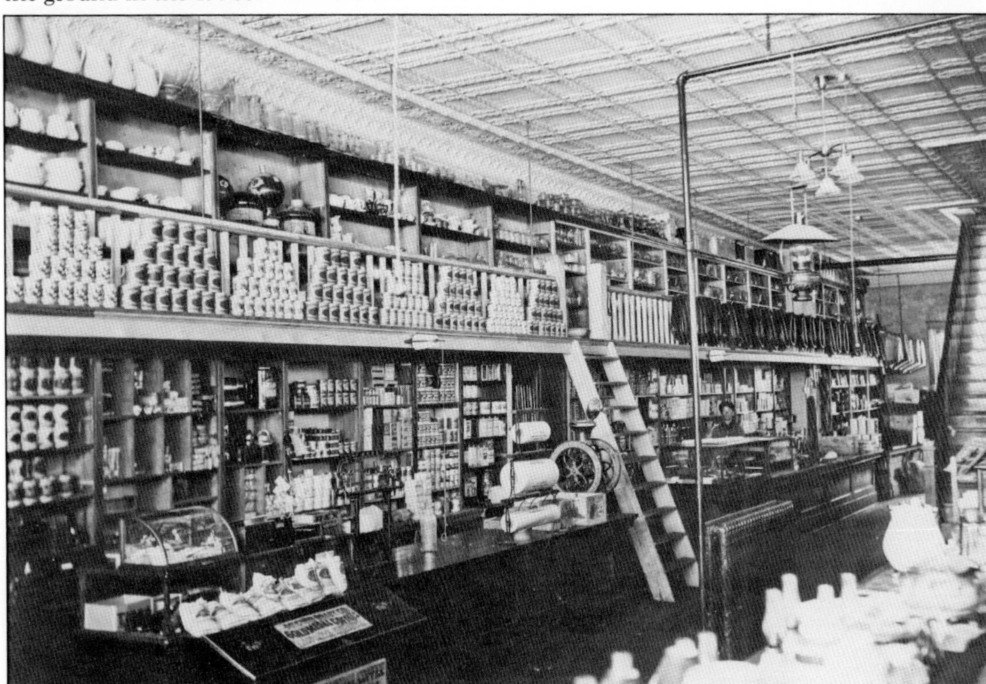

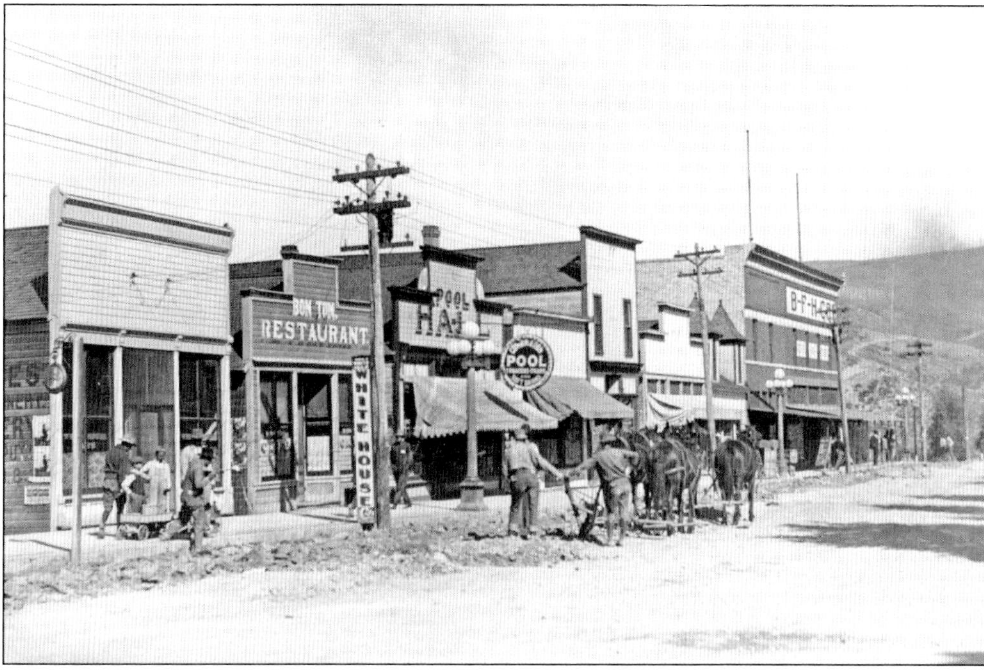

As the downtown area of Kemmerer grew, the town council was faced with the problem of sidewalks and paved streets. It was decided that every citizen would be assessed $2 to go into a fund to improve the condition of the downtown streets and sidewalks. If the person could not pay, he would have to work one day on the streets. By 1935, Kemmerer had paved streets and good sidewalks. Winter brought a whole new problem to Kemmerer. The snow was removed from the mines using a plow on the front of a railroad helper engine, but there was no way to remove the snow from the Kemmerer streets. Most people would use a horse and sleigh or walk in the winter months.

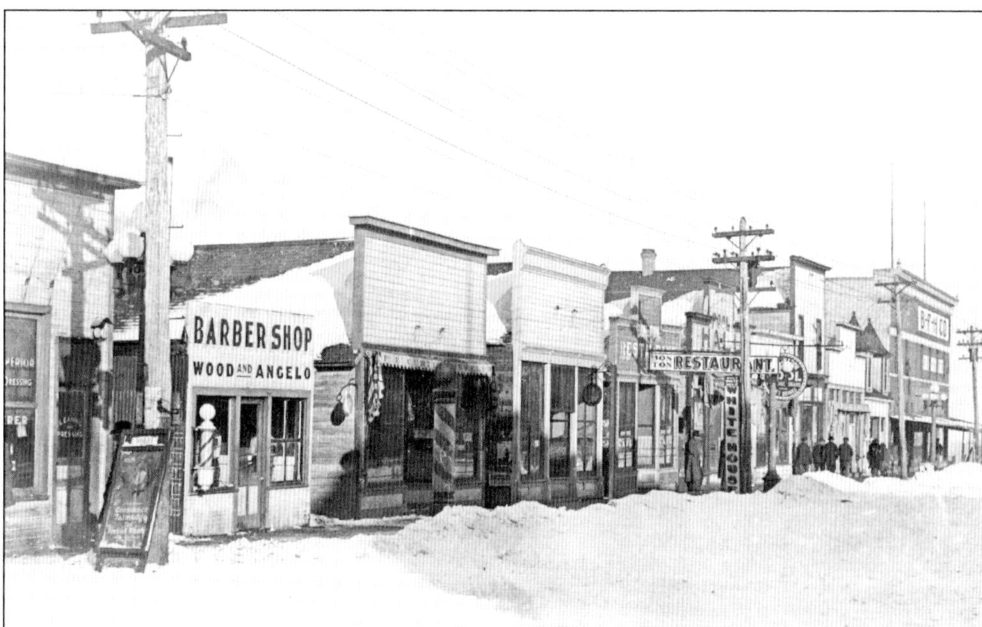

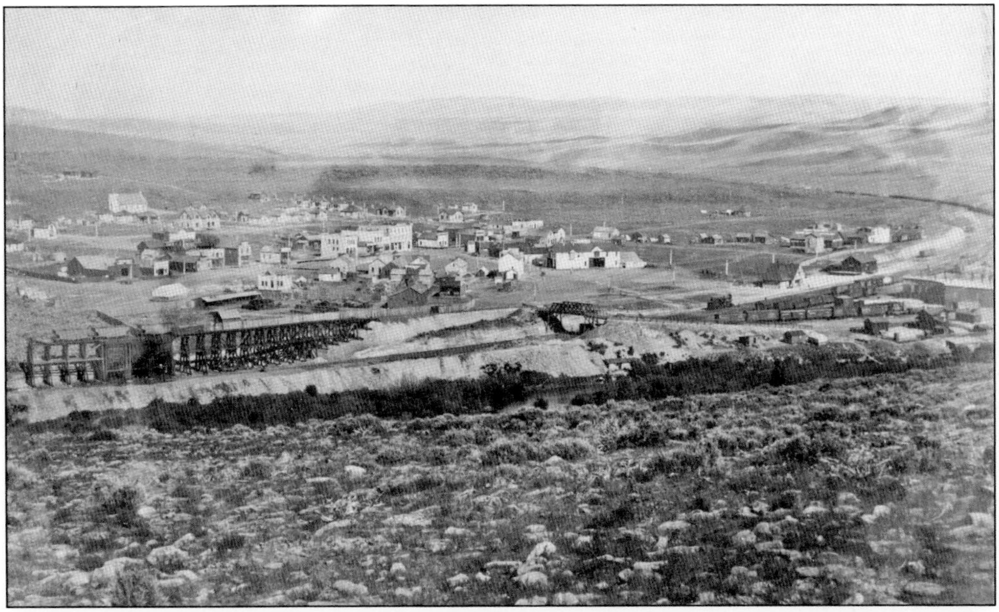

In the spring of 1897, Quealy and Kemmerer informed the Uinta County Commissioners that an overhead bridge was needed in Kemmerer. During the spring runoff, it was not safe for people to cross the river to get into Kemmerer. In 1898, a bridge was completed; this bridge (above, c. 1911) went over the river and railroad tracks and ended in front of the Frontier Supply Company store. The overpass was used until May 1936, when a new bridge located on the north end of Kemmerer was built. This bridge (below, c. 1980) also goes over the river and the railroad tracks. A pedestrian walk was built next to the new bridge so people from the Flats could safely walk into Kemmerer. The new bridge is still in use today.

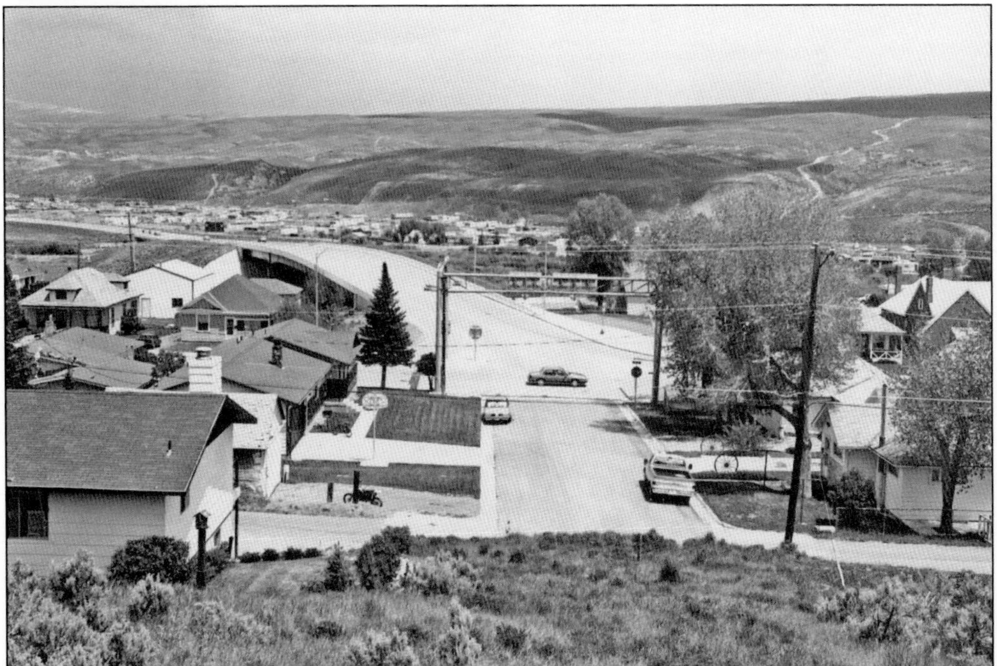

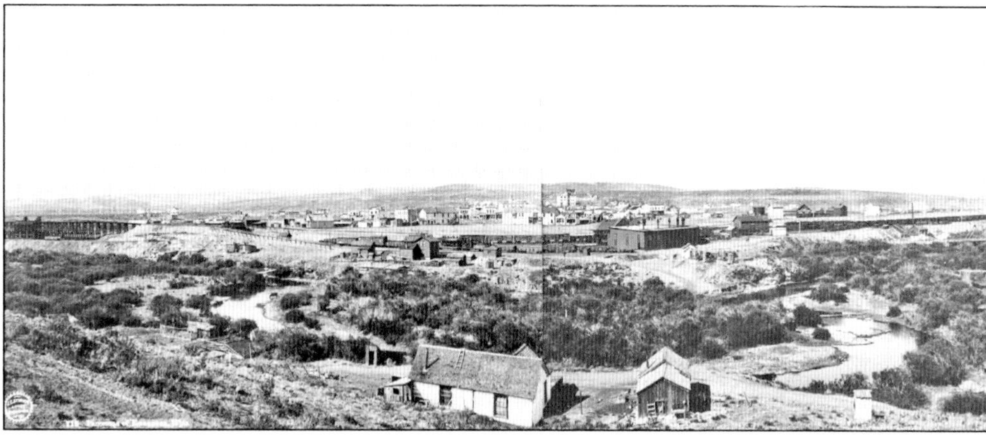

In the fall of 1897, when the town of Kemmerer was established, there were only a few large buildings (above): Keenan's dance hall, the Elk Hotel, and the Kemmerer Hotel, which was just beginning to be constructed. By the summer of 1911 (below), the town had grown significantly. Bitter-cold winters, summer droughts, fires, and mining accidents have left their mark on Kemmerer. But despite everything, the town of Kemmerer and the people who call it home have continued to thrive and can rightfully claim that they have accomplished their task in the wilderness. Kemmerer was not built by a single institution but by a whole community.

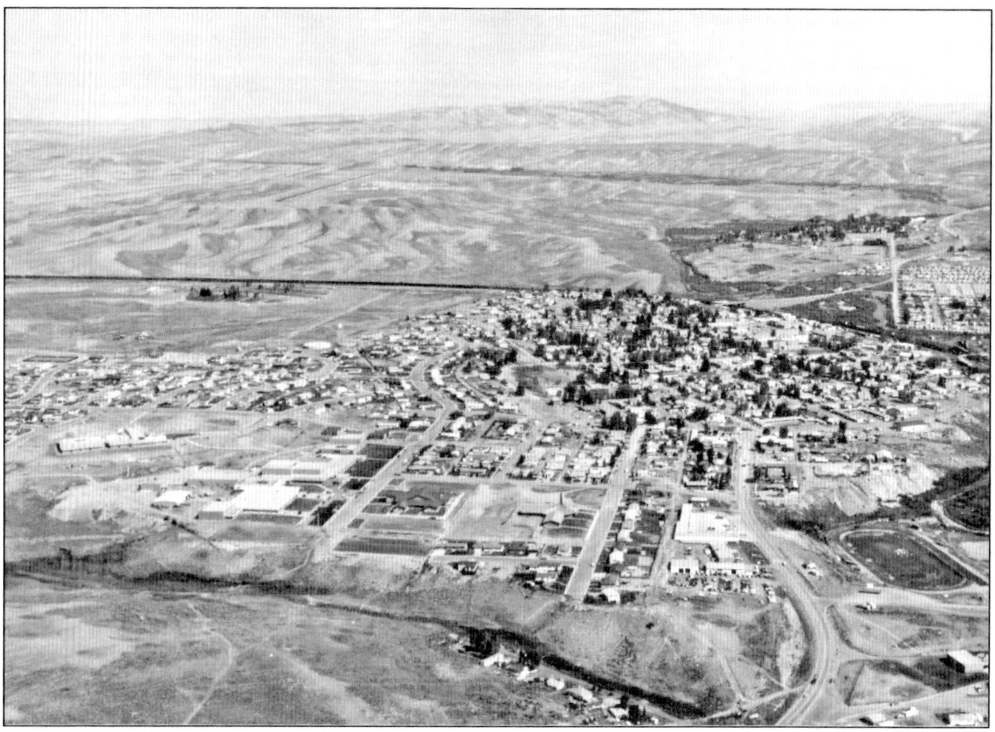

Four

Moonshining

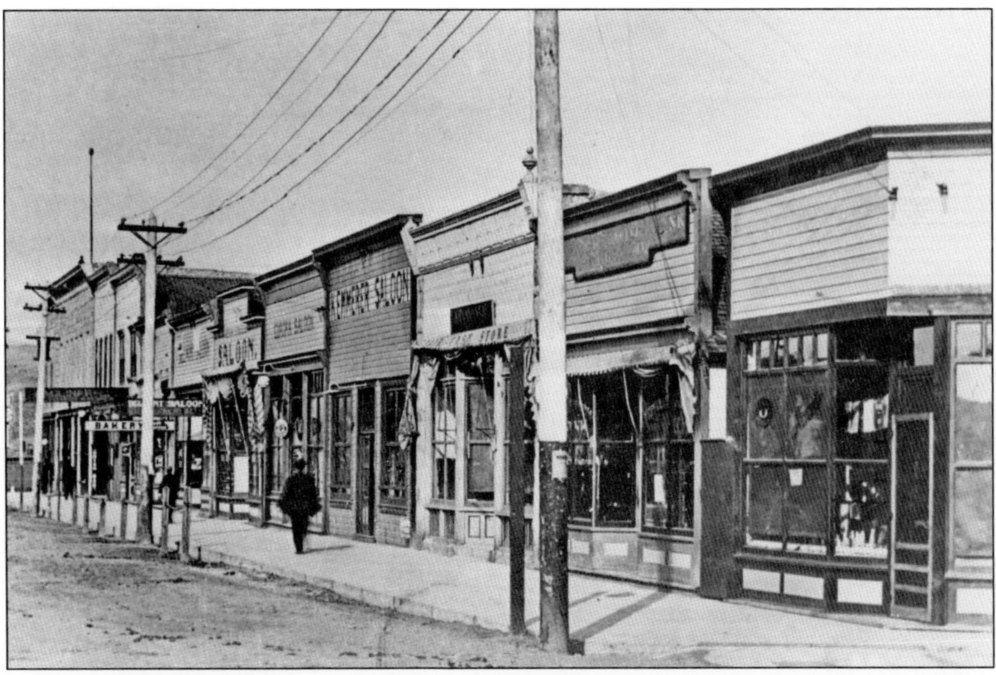

Rows of saloons lined the Triangle Park in 1918; in just one short year, the 18th Amendment to the Constitution prohibiting liquor was passed. On June 30, 1919, crowds gathered along the Triangle to witness the passing of "John Barleycorn" and all that the name symbolized. A band played music while a hearse was escorted by pall-bearers to the grave site on Klondyke Street. At midnight, all saloon doors closed.

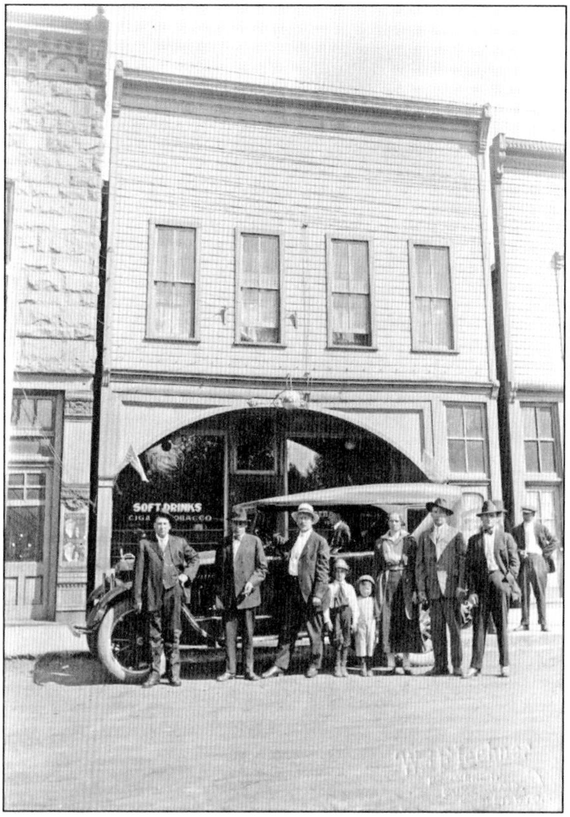

The Union Saloon (left) was built about 1905. Prohibition brought great changes to the local saloons. The Union Saloon changed into a soft drink and pool parlor. The drinking establishments were no longer permitted to serve alcohol, although many still did. Many of them opened as soft drink or pool parlors, converting bartenders into soda jerks. A few of the saloons went out of business. The only bar in Kemmerer to survive and never change its name during Prohibition is the Stock Exchange Bar (below, 4th from right). It was built about 1905 by Peter Gordon. During Prohibition, it was changed into a men's club, but liquor was served in the basement. Customers could bring their bottles into the club and the bottles would be sent to the basement to be filled. The bar, still in operation, is owned by the Julian family.

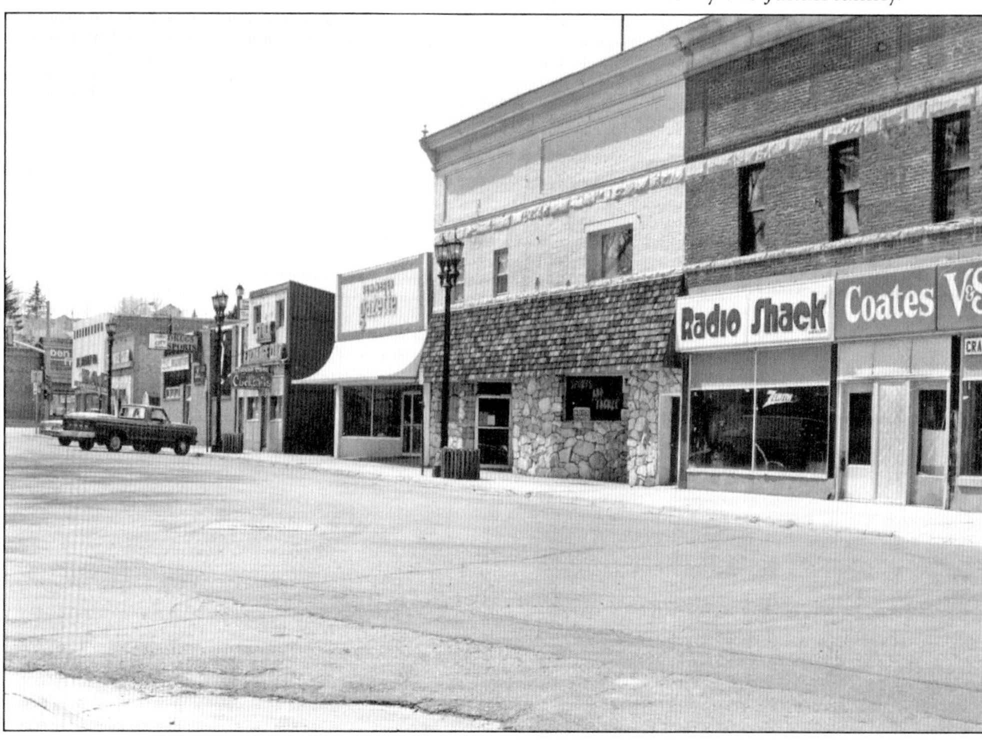

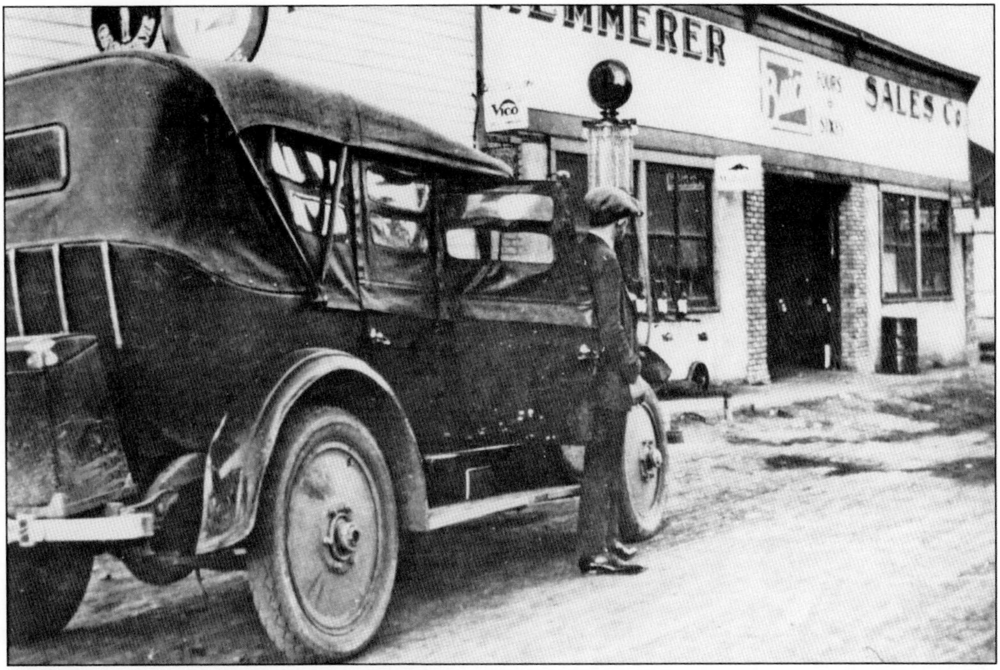

Prohibition did not stop people from buying or consuming liquor. As the demand for liquor grew, Kemmerer and Diamondville became national centers for making and supplying liquor for the Midwest. Local men would work their jobs during the day and tend to the moonshine at night. Large quantities of moonshine were produced in the area, but little was consumed by the bootleggers. Both manufacturing and transporting were profitable illegal operations. Kegs, sugar, and other supplies arrived almost daily, and shipments left with equal frequency. Most shipments left in automobiles that had been taken to the local garages and outfitted with heavy-duty springs so they could haul large quantities of liquor.

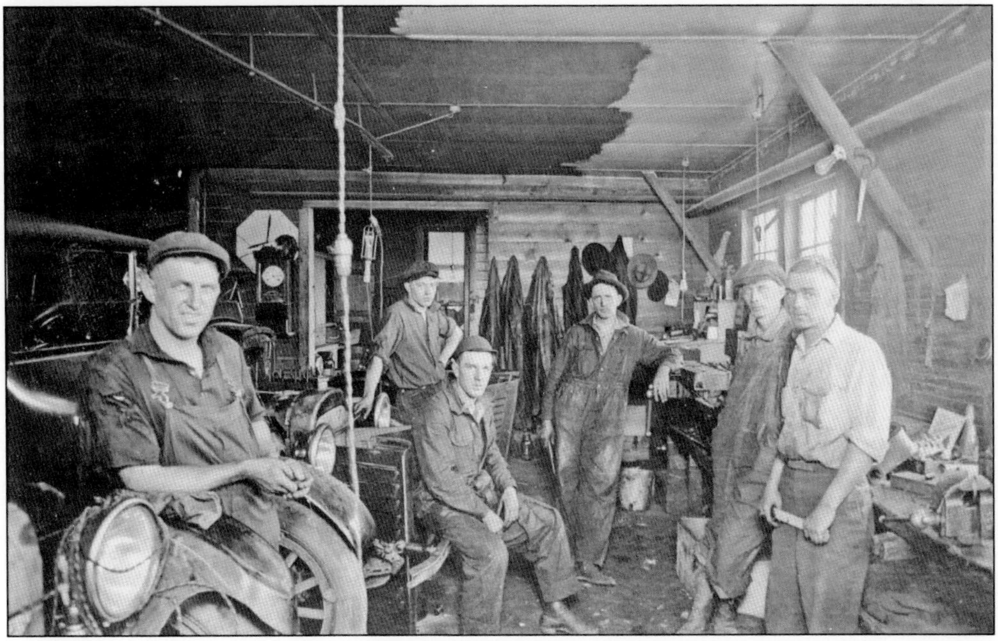

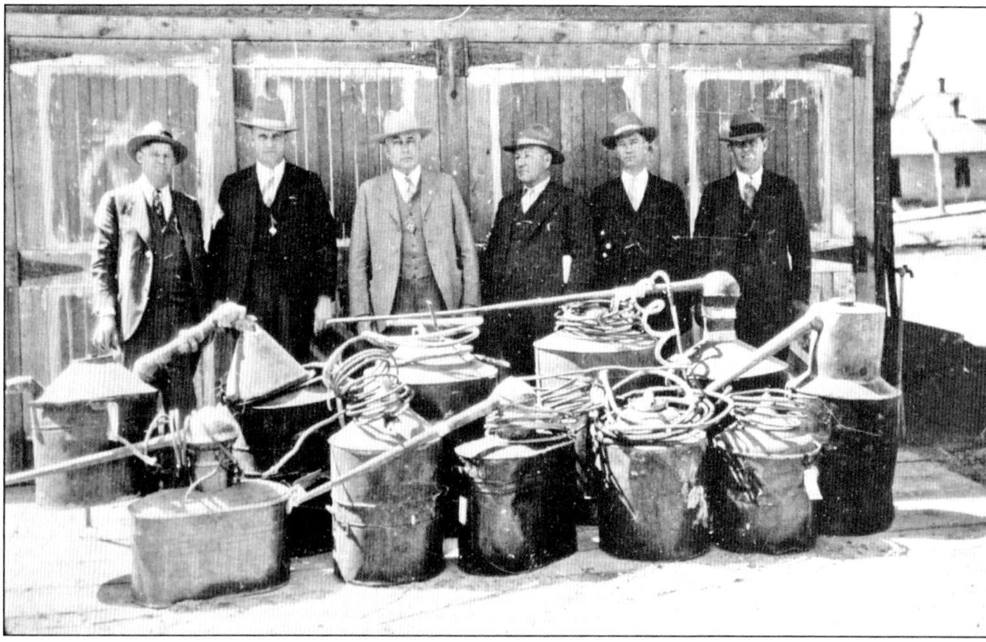

On March 8, 1920, five barrels of wine were poured into the sewer through an opening in the floor of the basement in Kemmerer's town hall. This was the first confiscated liquor destroyed in the Kemmerer and Diamondville area. However, 20 cases of liquor were found near the Hams Fork River a month earlier but were not destroyed until the owner was found guilty. In one raid, the revenuers found a still mounted on the back of an automobile. The bootlegger was trying to hide the car and the still in the sagebrush. When the hideouts were found, the stills were confiscated and the liquor destroyed.

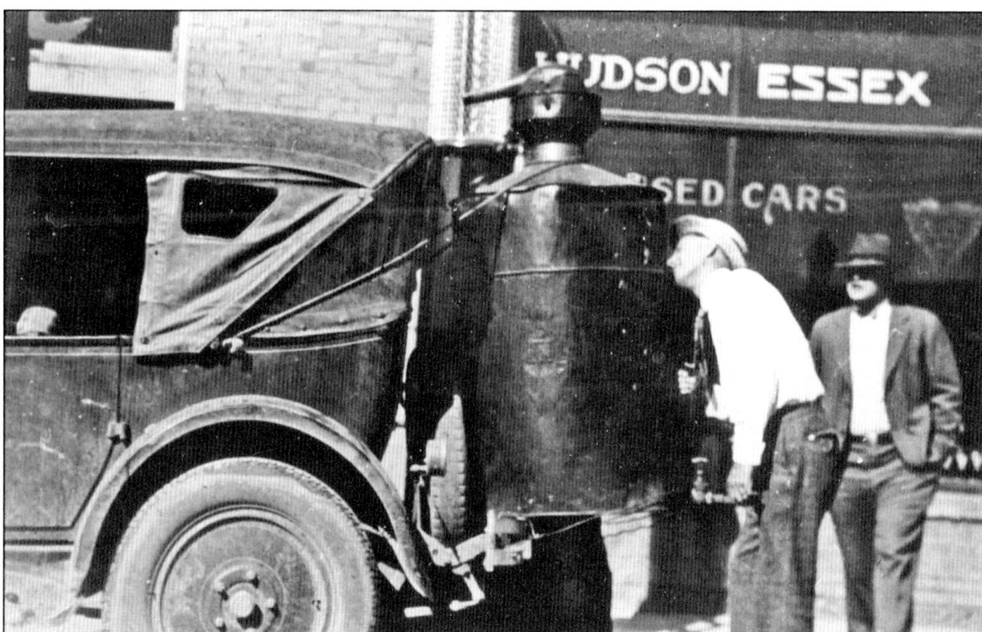

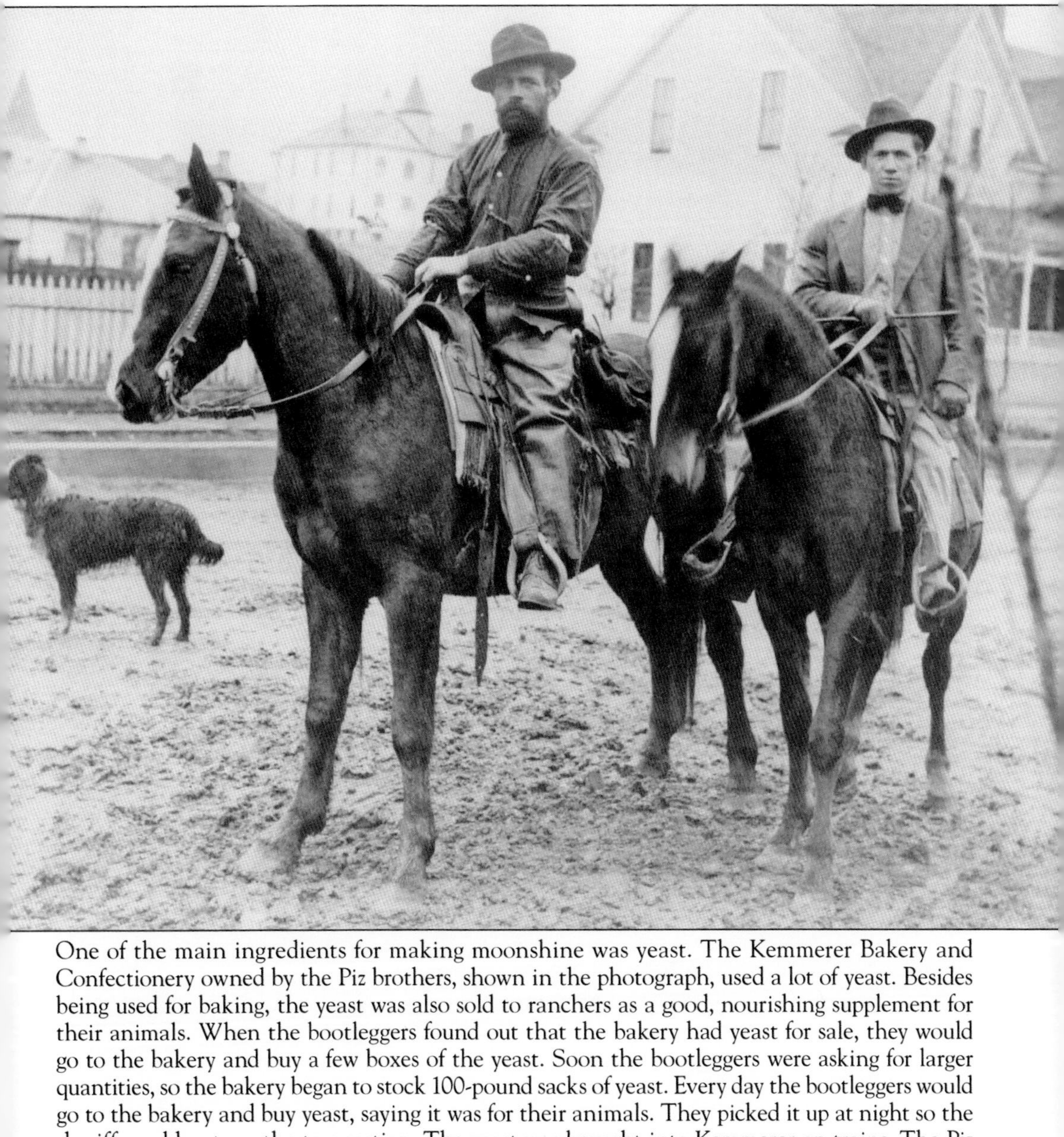

One of the main ingredients for making moonshine was yeast. The Kemmerer Bakery and Confectionery owned by the Piz brothers, shown in the photograph, used a lot of yeast. Besides being used for baking, the yeast was also sold to ranchers as a good, nourishing supplement for their animals. When the bootleggers found out that the bakery had yeast for sale, they would go to the bakery and buy a few boxes of the yeast. Soon the bootleggers were asking for larger quantities, so the bakery began to stock 100-pound sacks of yeast. Every day the bootleggers would go to the bakery and buy yeast, saying it was for their animals. They picked it up at night so the sheriff would not see the transaction. The yeast was brought into Kemmerer on trains. The Piz family continued to supply yeast to the bootleggers until 1933 when Prohibition was repealed.

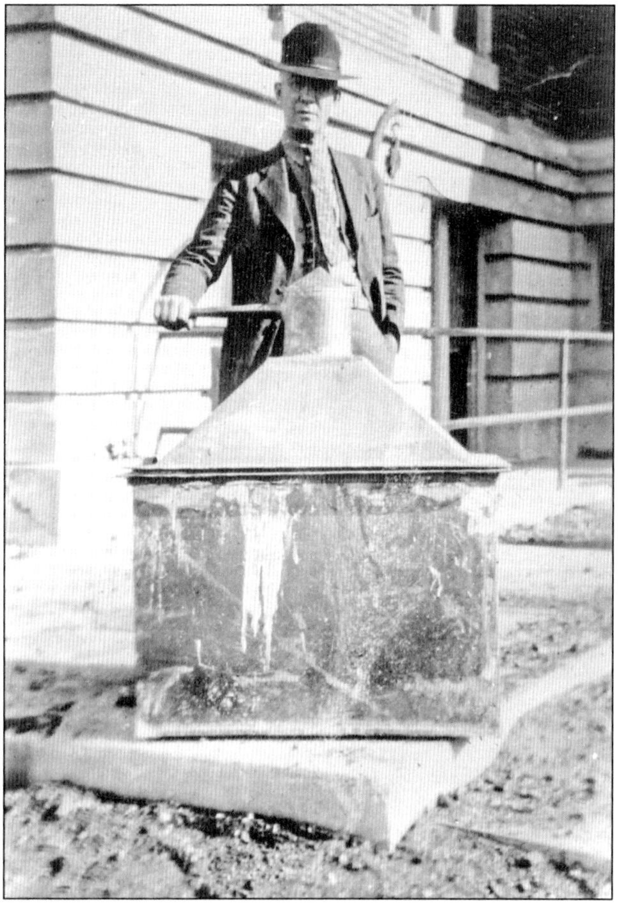

During the 1920s, the confiscated stills were taken to the Lincoln County jail to be destroyed. Stills were being built behind locked doors, and the revenuers could not keep up with the stills or the bootleggers. When Prohibition began, everyone was expected to dry up. This resulted in those who had never indulged before having the desire to do so. Demand for liquor grew with every passing day. Law enforcement agencies soon realized they could not effectively enforce the law. It was an unsavory job, and there was a constant turnover. There were a few revenuers who would let the bootleggers out of jail at night so they could take care of their moonshine, but they would always return to the jail before daylight.

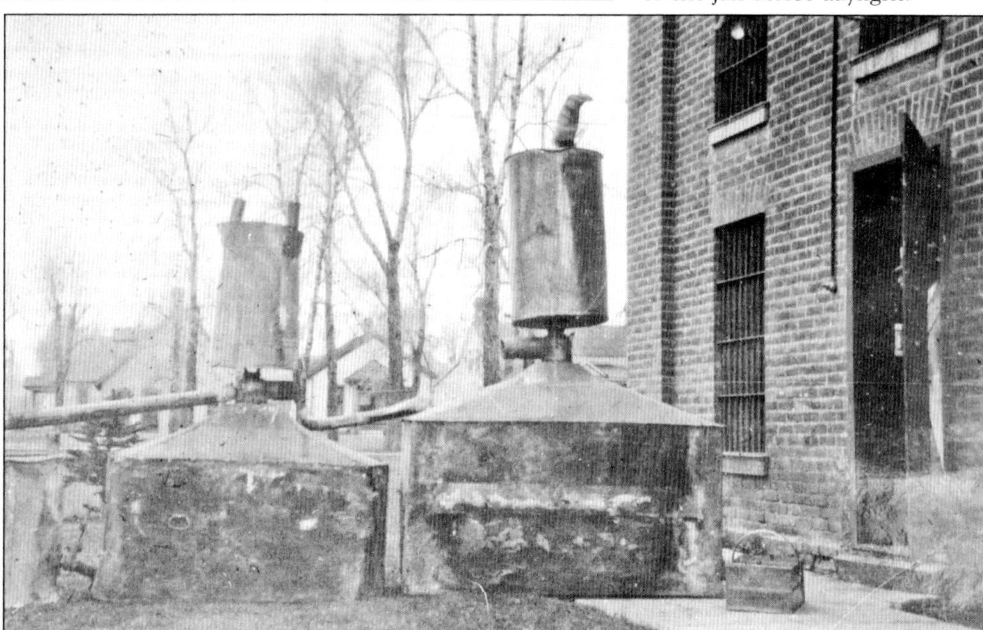

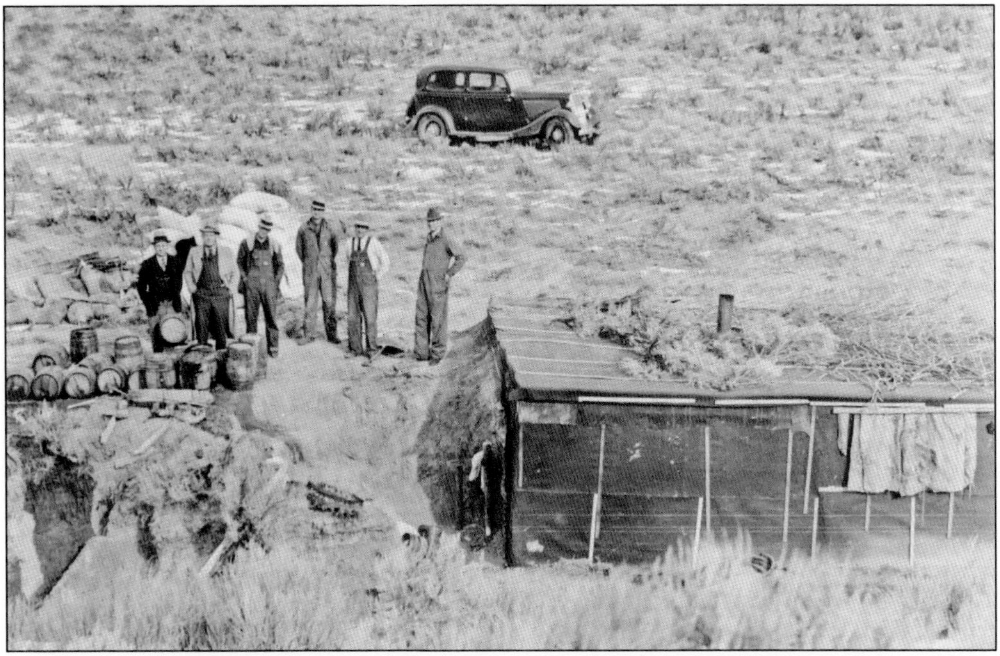

In 1931, the Kemmerer area saw a large number of raids. There were nine total revenuers in Kemmerer; two of them were permanently assigned here. For 12 years, Prohibition enforcement in the Kemmerer area had for the most part been disregarded. In 1931, well-publicized raids and destruction of the booze became the new rage. Along with the increase in raids also came the increase in violence. During a raid shown here in 1932, two officers seized a still, ingredients to make moonshine, and 6 gallons of whisky. The bootlegger asked to take care of his animals but instead grabbed his gun, shot, and fatally wounded one of the agents; the bootlegger was also killed. Federal agents returned to the bootlegger's hideout and burned it along with the moonshine.

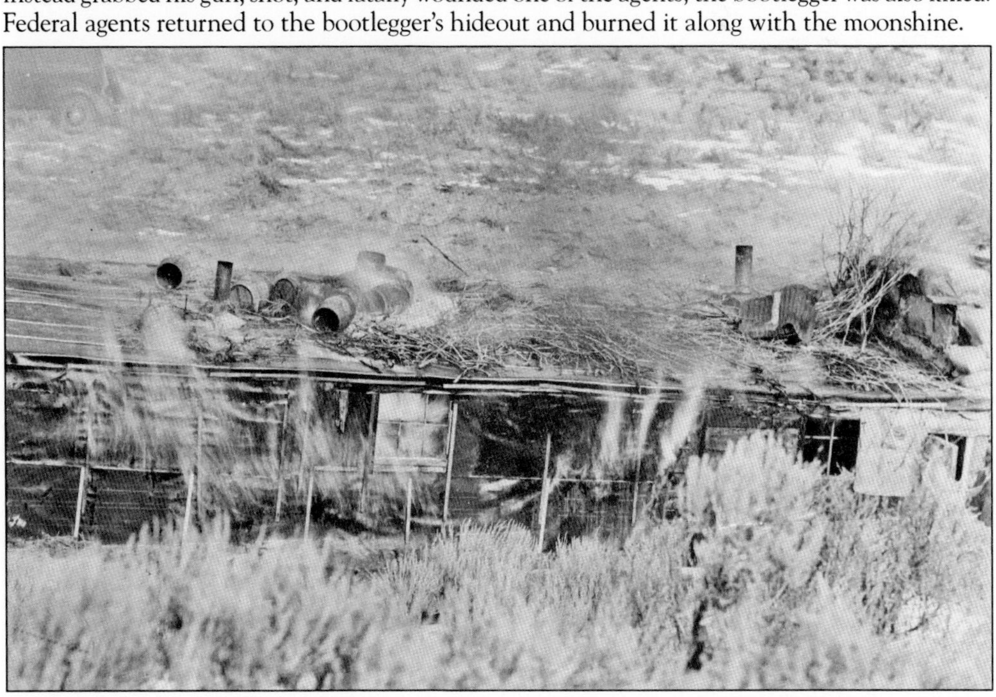

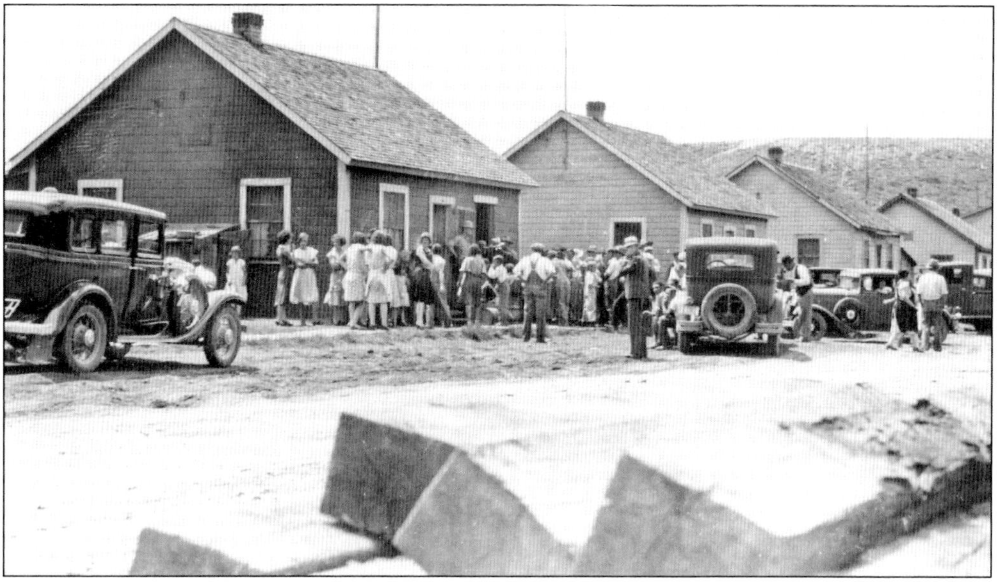

Bootlegging was not only a rebellious act; it became a way of life. Those involved in the illegal activity took great pride in producing a high-quality drink. The production and transportation of the moonshine was very profitable and contributed significantly to the area's economy. Women and children played essential roles in the bootlegging operation. It was quite common for women to be drivers because they were the least suspected to be hauling moonshine. Children would be given candy by the revenuers so they would reveal locations of stills and names of the bootleggers. The children would accept the bribes and send the revenuers on wild goose hunts looking for nonexistent stills. Shown here on June 25, 1931, agents discovered a huge cache of moonshine consisting of 13,000 gallons in 10-gallon kegs. The liquor was stored in five vacant houses.

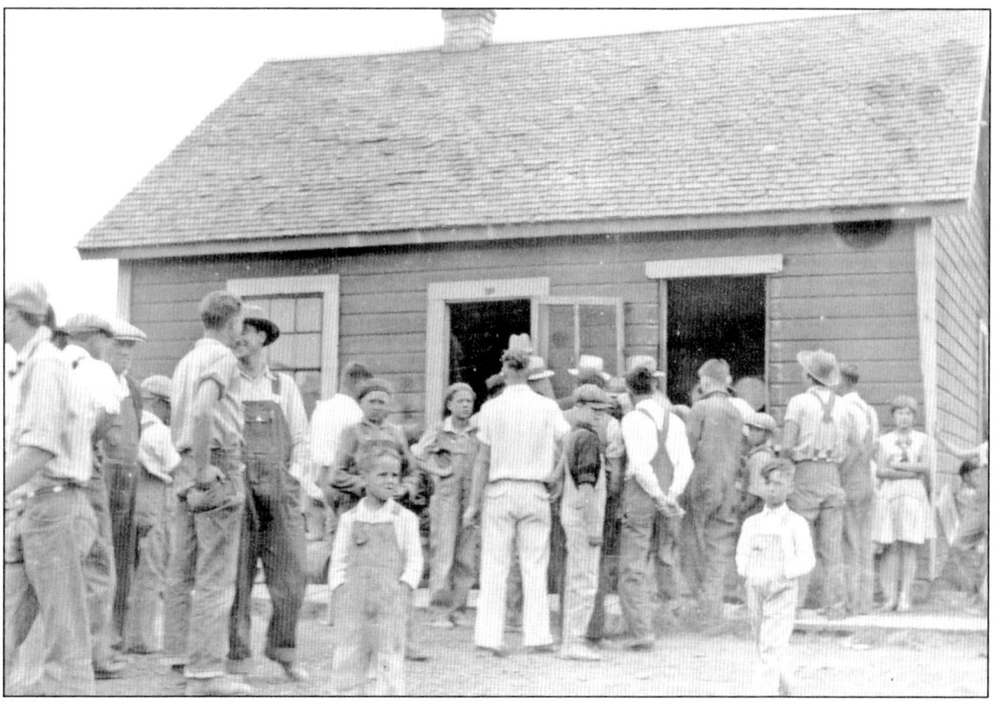

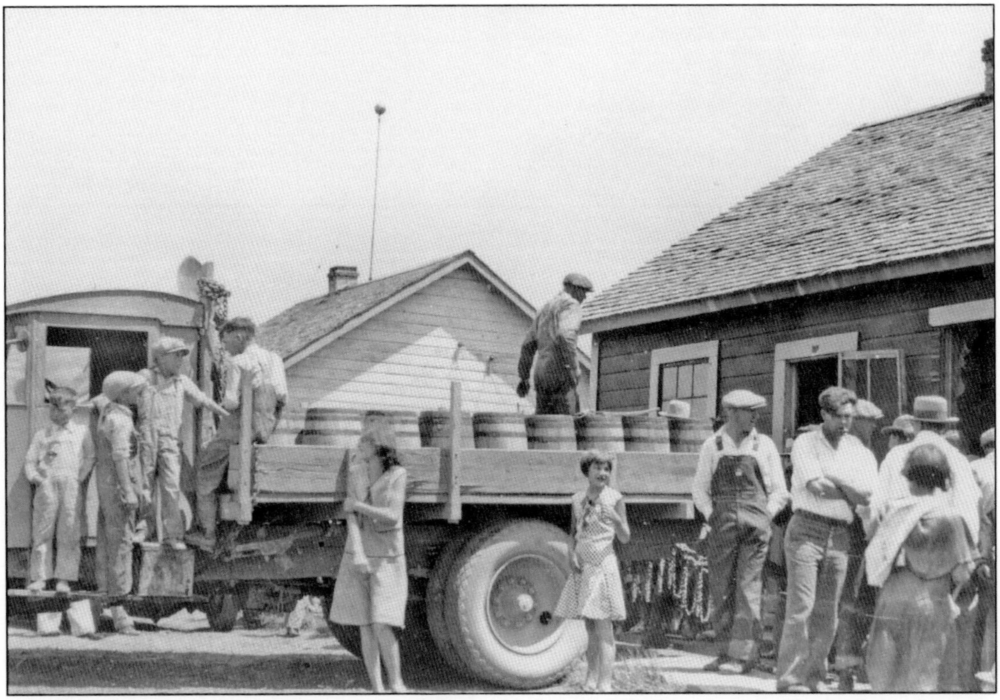

As the revenue officers prepare to haul away the confiscated moonshine, a crowd gathers to watch as the kegs are smashed and the contents are poured into the gutters. Some mourned the loss, while others seized the opportunity to collect the precious drink. The kegs were hauled out of town to be destroyed. Had the moonshine been sold, it would have netted the owner over $90,000. Because the feds did not work at night, the trucks they hired during the day to haul the confiscated moonshine were hired by the bootleggers at night to haul the moonshine out of town, where they hid it in gullies and covered it with sagebrush.

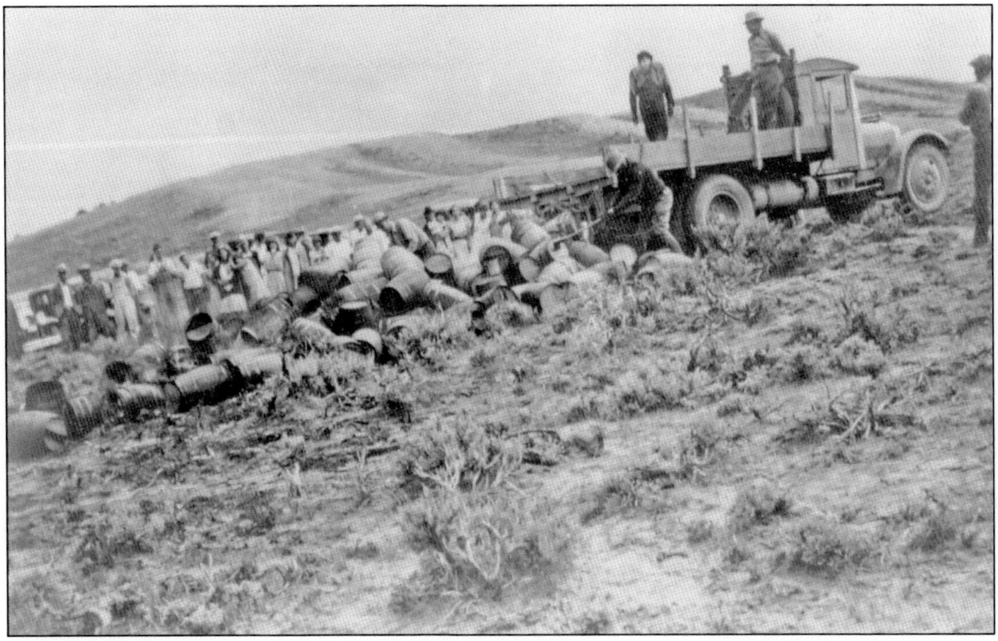

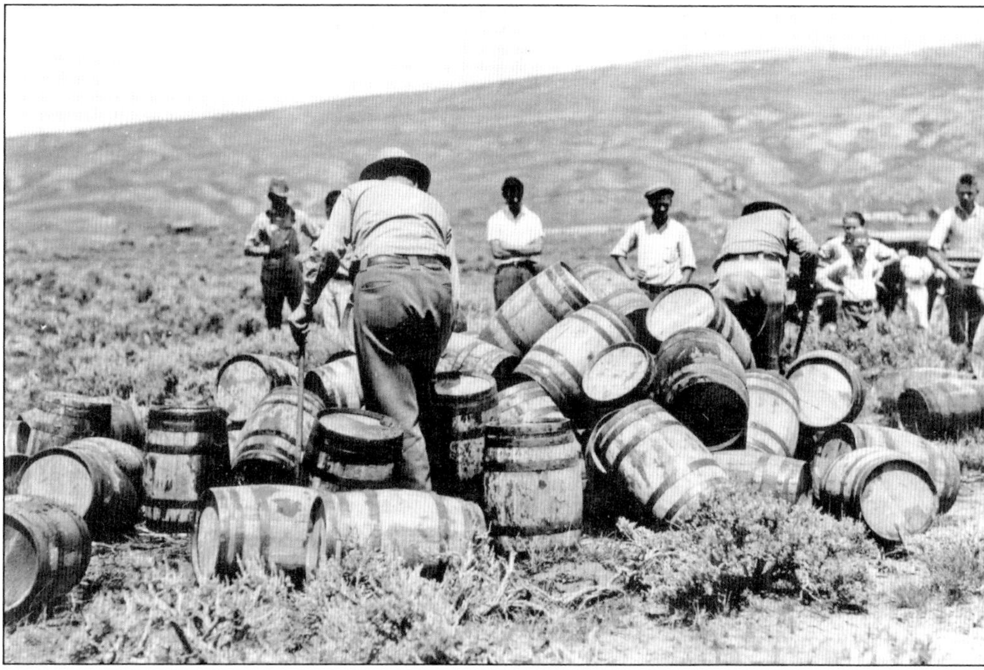

By early 1931, federal agents determined that almost all of the confiscated moonshine came from Kemmerer. The moonshine business by this time was a well-run, tightly organized industry. The government demanded that Wyoming authorities take action. The largest mass destruction of moonshine occurred on October 24, 1931. The moonshine was taken to an area just south of Kemmerer and burned. An impressive fire burned for over an hour and could be seen for miles. Those involved with enforcing Prohibition became quite creative with ways to destroy the liquor. In 1933, the 21st Amendment to the Constitution was passed, repealing Prohibition; this ended the long dry spell. John Barleycorn was exhumed from his grave, and bars were allowed to sell liquor legally. For many years following the repeal, the most popular drink across the country continued to be Kemmerer Moon.

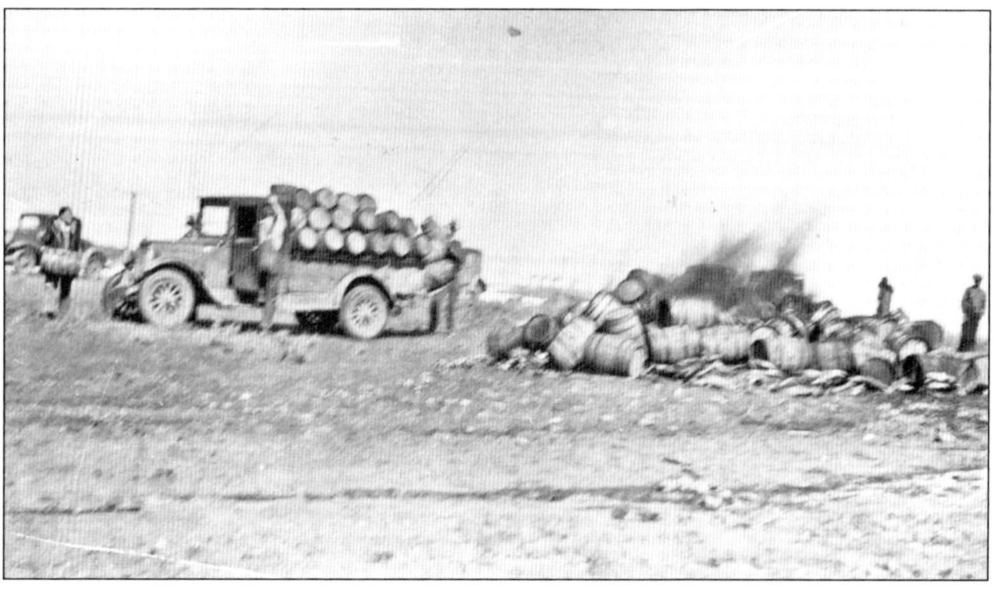

Five

Schools, Churches, and Businesses

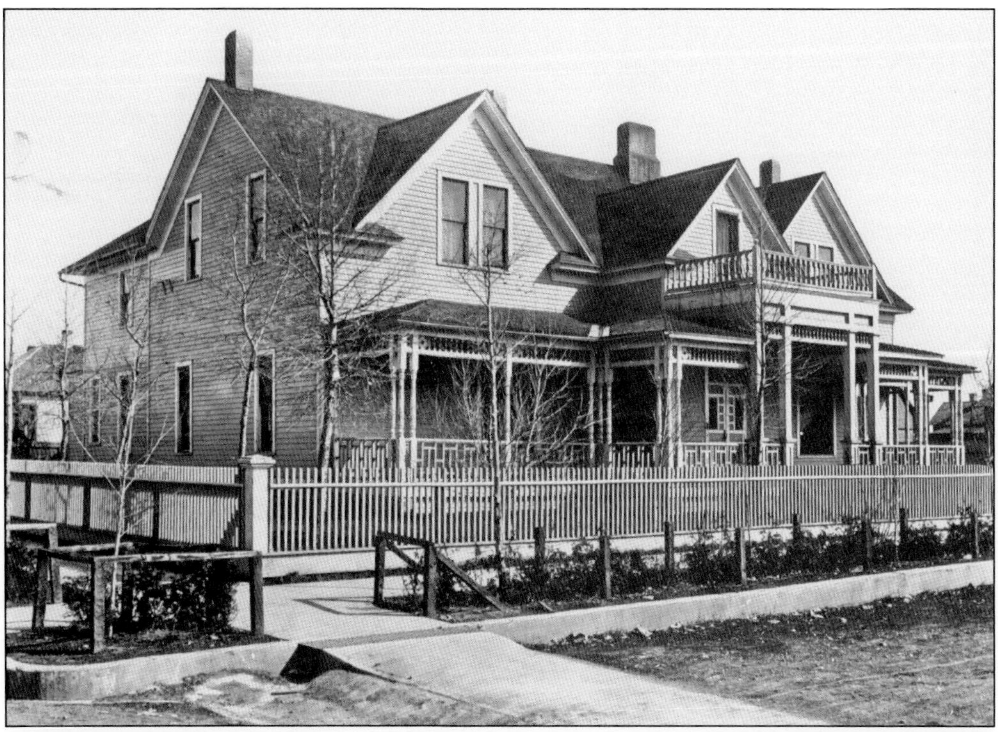

As with most growing communities in the small towns of the West, schools, churches, and businesses always followed. Kemmerer's first school was located in the Quealy home. But some records indicate the first classes were held in the Sim Westfall building, the location of which is unknown. The school opened January 11, 1899; school was held in the Quealy home for a little over one year.

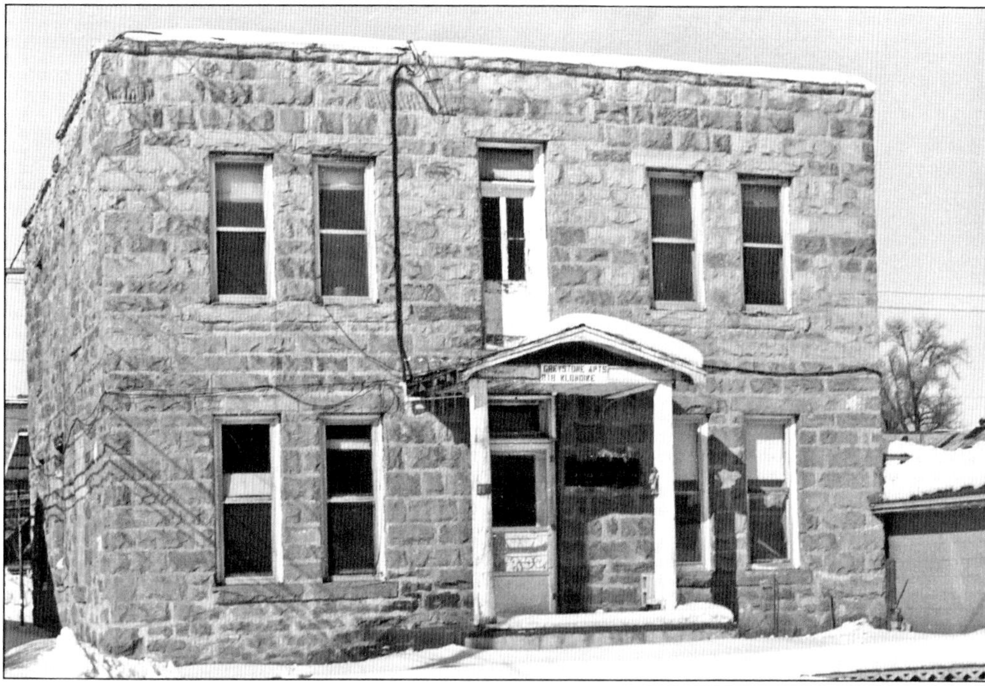

A new school opened in September 1902. The new school had around 45 students on the first day. The school had four classrooms, but two of the rooms were left unfinished. The first teacher to come to the area was Grace King. The new school was in need of a school bell, so the Kemmerer parents, along with the Quealys, rallied to help the school. This was the early version of the PTA; there was no formal organization for many years. The students performed a fund-raising play held at the Casey Grand Opera House.

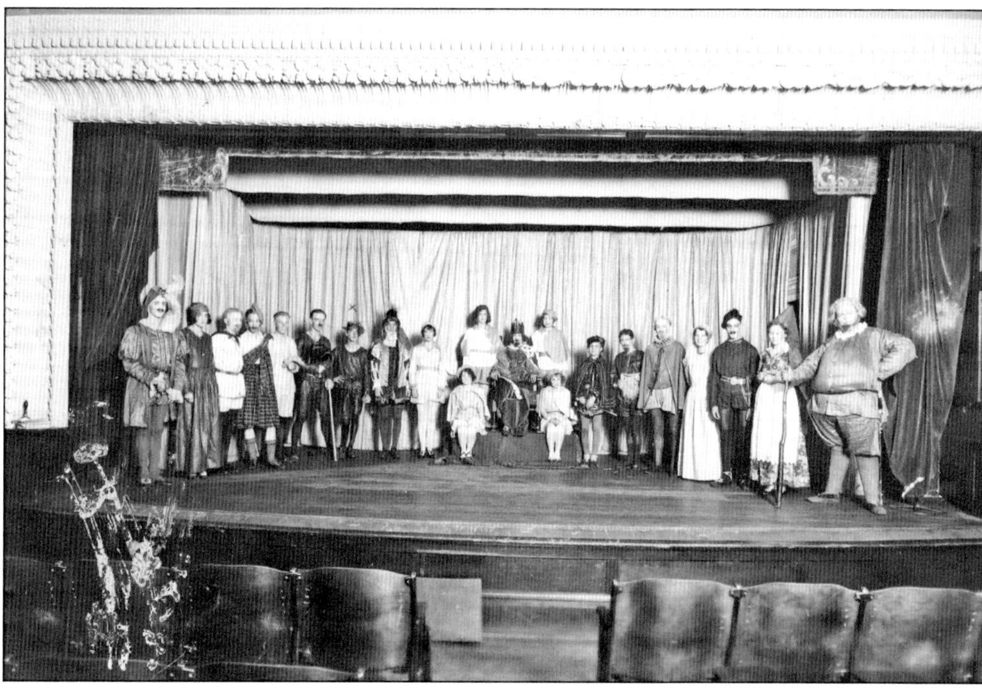

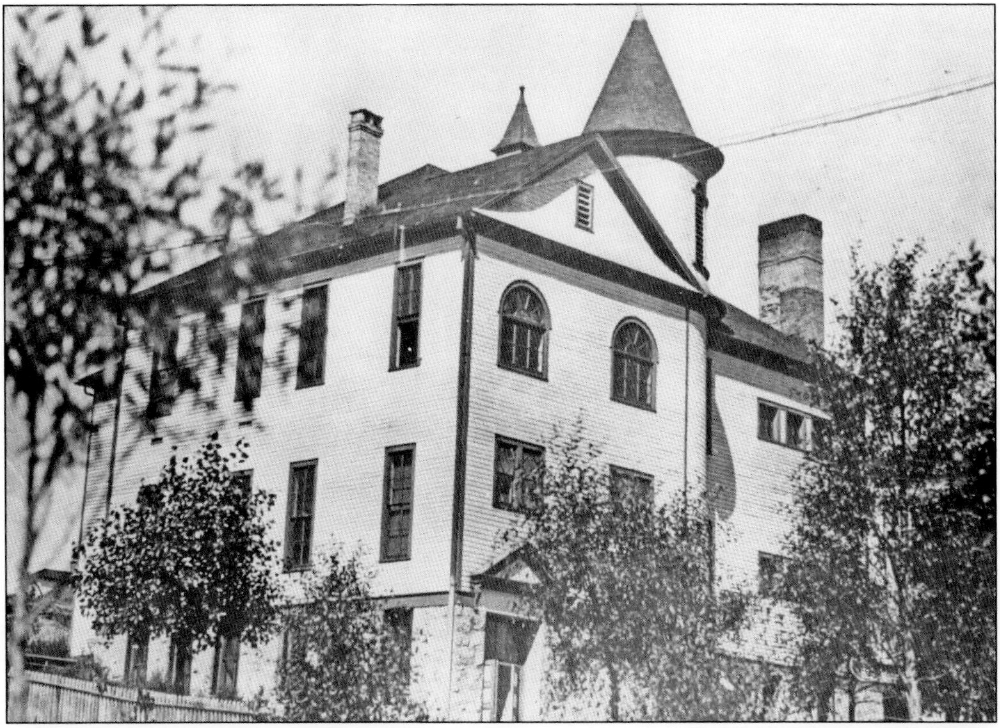

As the population of Kemmerer increased, so did the need for yet a larger school. A larger building (above) was constructed in 1910, and students entered the new school in the fall of 1911. One hundred students registered on the first day of school; at this point, only a few were high school age. Most of the high school students were girls, as most of the older boys were working in the mines. The teachers were men or unmarried women. The unmarried teachers would board with a respectable family in Kemmerer. If the teacher got married, she would be let go. The belief was that the husband made enough money to support the family. Below is a class photograph from the new school, date unknown.

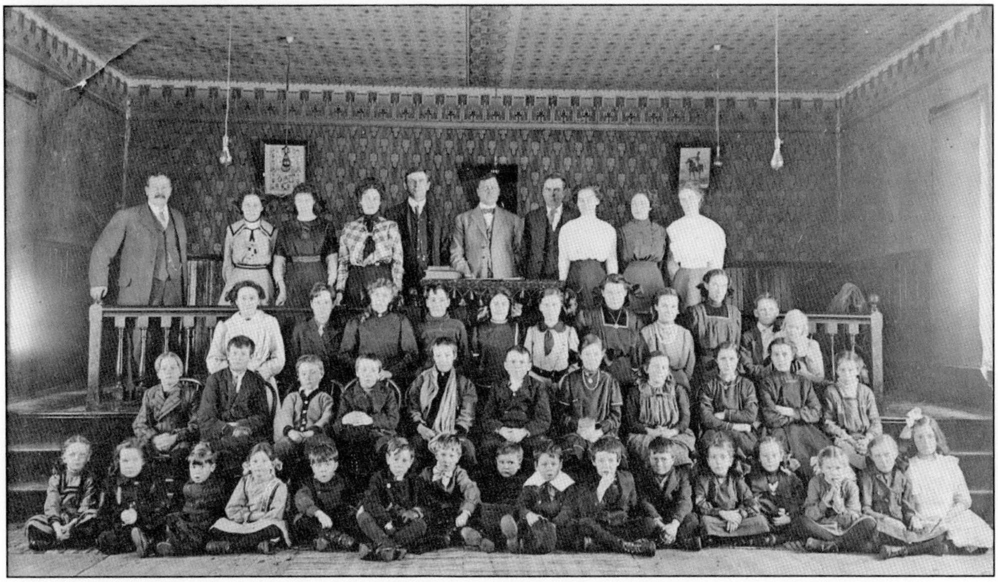

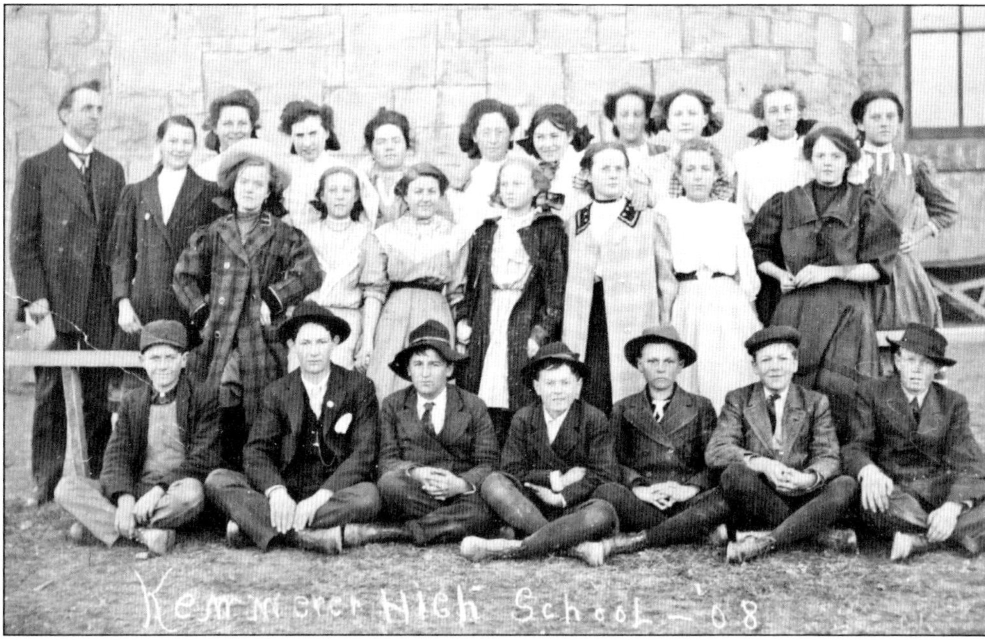

In 1911, the new school opened as Kemmerer Grade School or Central Grade School. The first senior class to graduate from the new school in the spring of 1912 had started their high school classes in 1908. At the beginning, the class had 24 students, but by the time they were seniors, there were only 11 girls who graduated. The young men were working in the mines, or the families had moved from the area. This class also had the honor of being the largest class to graduate. The school's superintendent in 1911 was Joseph E. Burch; he was also the senior class teacher. The third-grade class to enter the new school was also quite large; there were more boys than girls in this class.

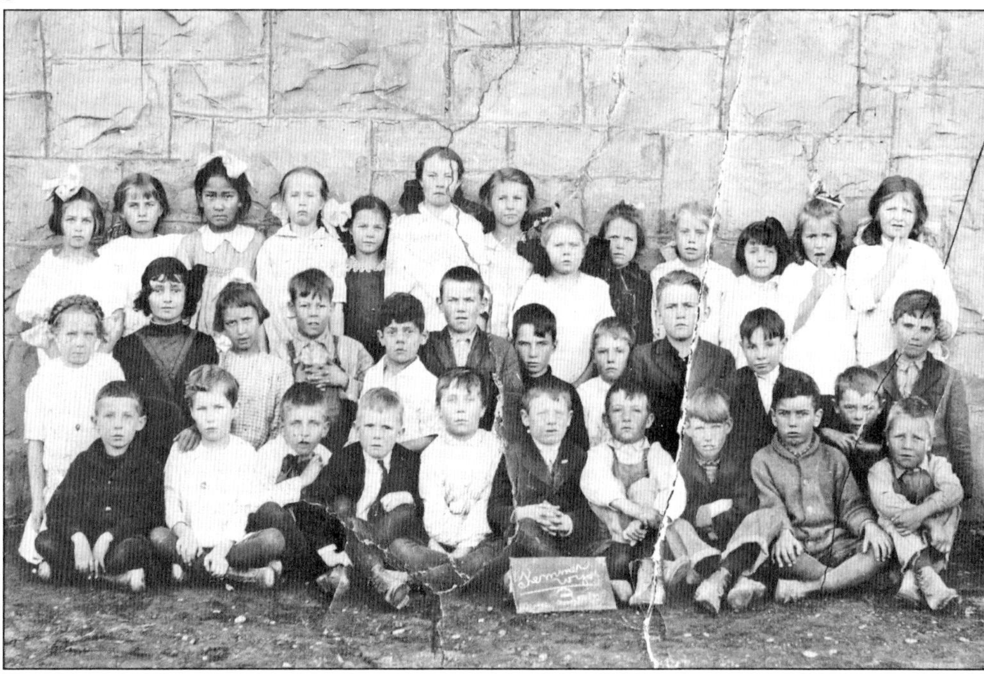

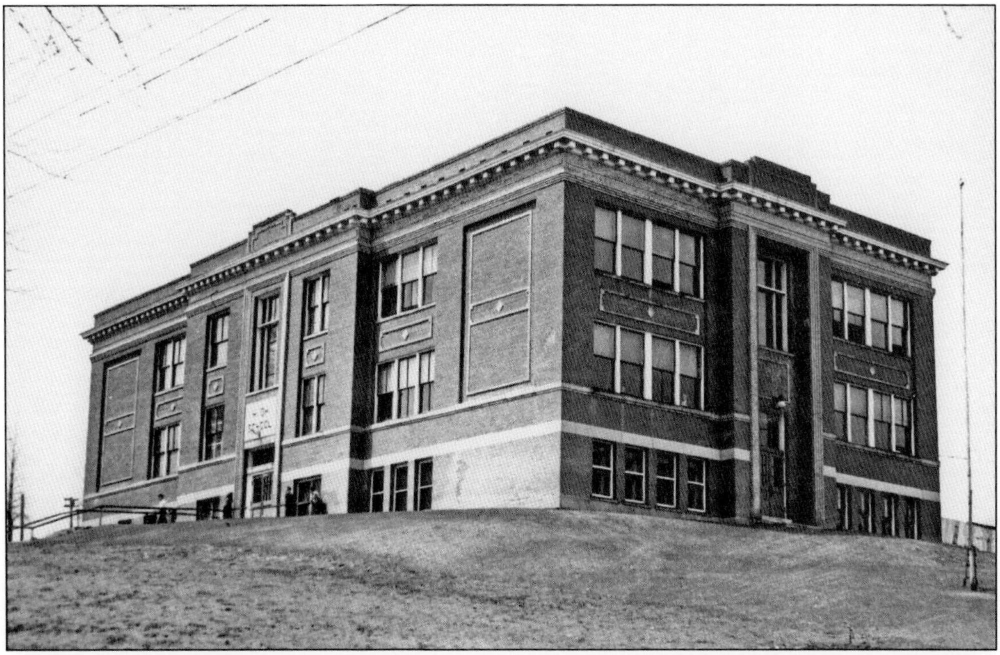

In 1912, the school board made a decision to build a high school (above). Kemmerer's first brick school was ready for students in the fall of 1913. There was not a gymnasium originally built in the school, but in 1920, one was added on. This building was used until 1963 and then torn down. The Lincoln Country Library was built where the old high school stood.

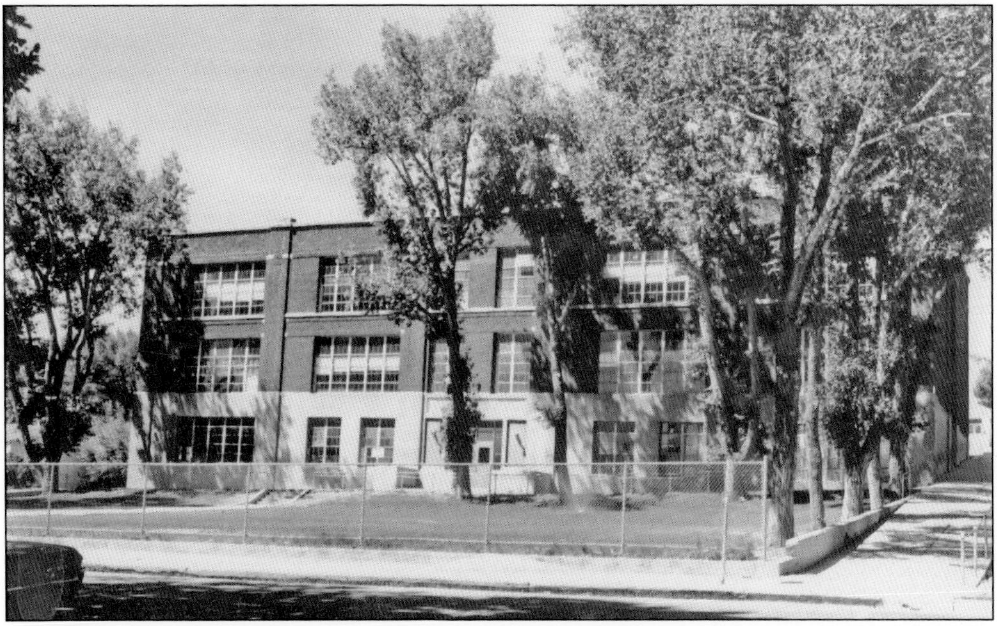

Over the years, the number of Kemmerer students steadily increased. In the fall of 1927, the elementary and junior high students attended a new school; this was Kemmerer's second brick schoolhouse. Students went to this school until 1972. The building was vacated and eventually torn down after a bond was passed to build an updated elementary. Emerald Housing was constructed on the old school yard.

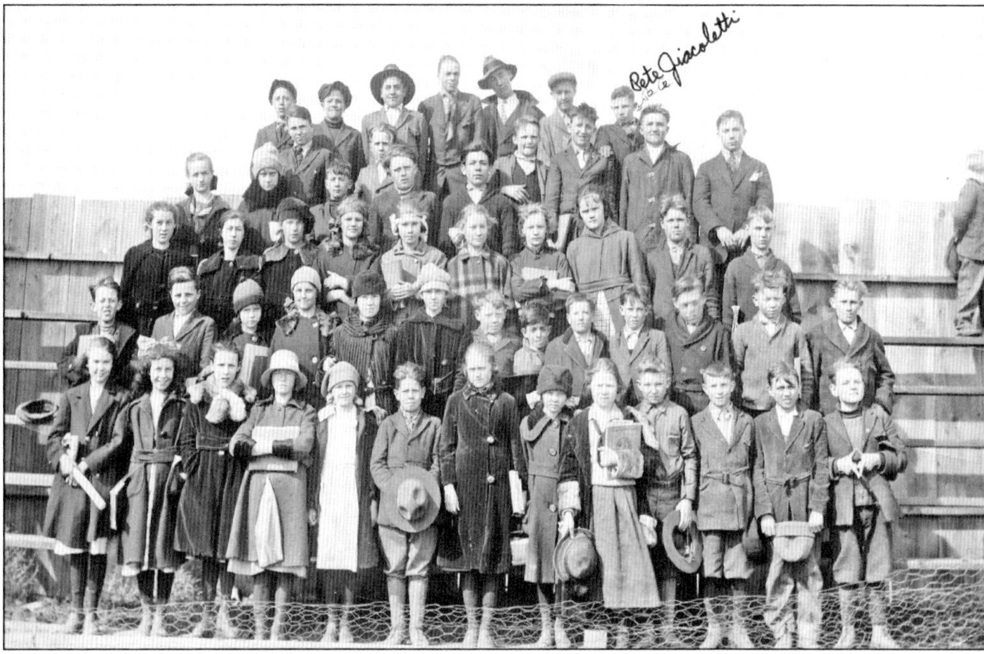

In the fall of 1919, the seventh- and eighth-grade classes (both pictured) almost doubled in size. The reason for this was that all the older students from the area coal camps had to finish school in Kemmerer. The camps had schools for the younger students but were not equipped for the older ones.

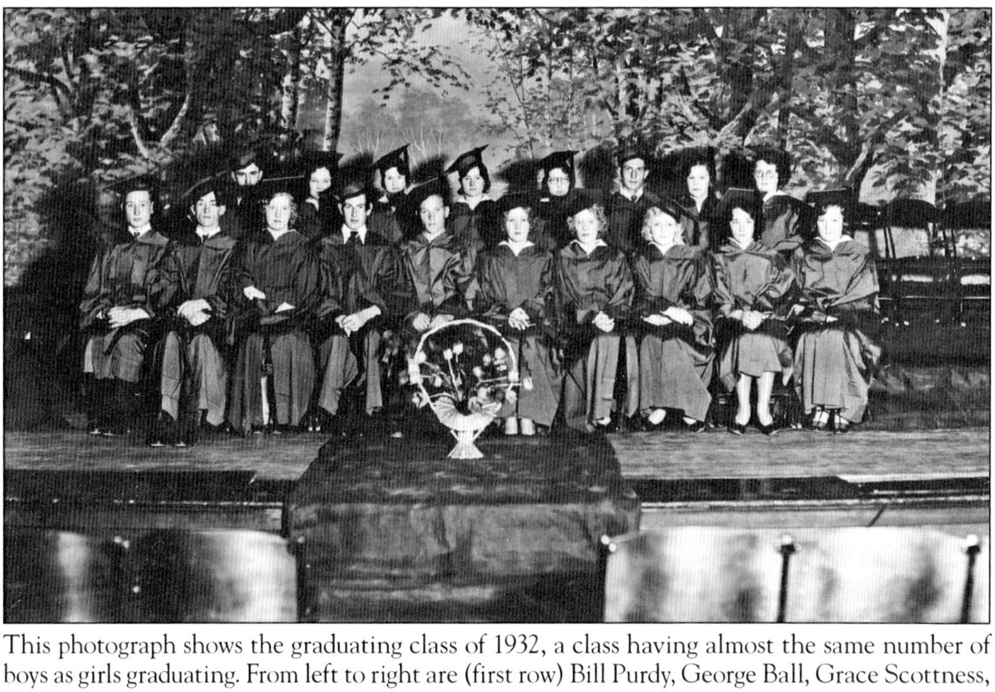

This photograph shows the graduating class of 1932, a class having almost the same number of boys as girls graduating. From left to right are (first row) Bill Purdy, George Ball, Grace Scottness, Robert Aken, Ralf Calvert, Meilme Hendrickson, M. Clark, Irene Kosky, Edith Ravenello, and Jessy Dobnick; (second row) Pete Casanova, unidentified, A. Bryant, unidentified, Joy Peterson, G. Fransconi, Velma Sawley, and Barbara Burgoon.

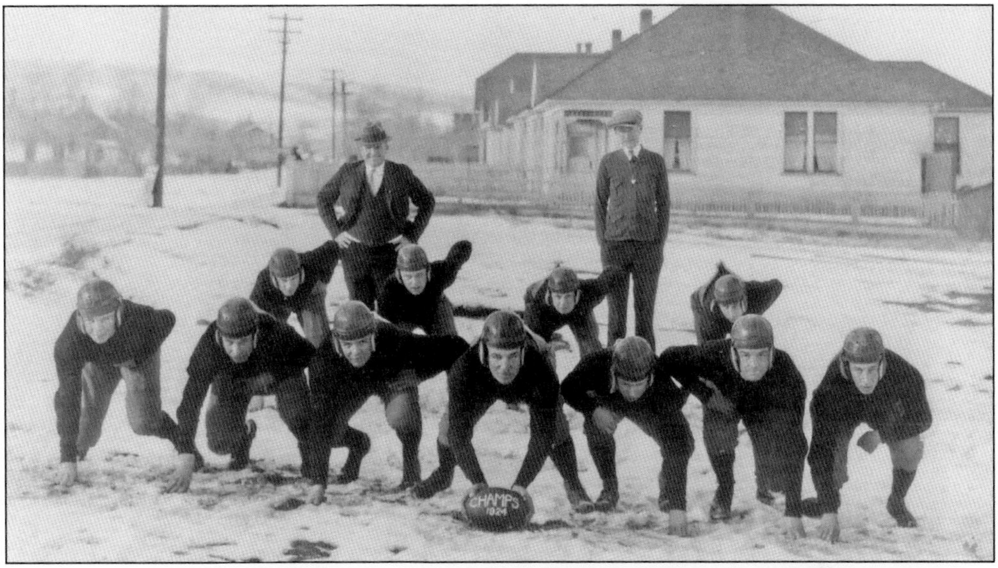

The first football game in the history of Kemmerer High School was played in 1923. The home games were played on land that is now the site of the courthouse. When the first game was played at Kemmerer, the people did not understand the game of football, and during halftime when the players went to rest, most people thought the game was over and left. The second half of the game was played with few spectators. The first game, played out of town in Evanston, was won by the Kemmerer Rangers. The remainder of the season was not as successful. But in 1924, the Kemmerer Rangers were crowned state champions. Basketball was also a favorite high school sport. In 1926, the high school teams for both boys and girls won state titles.

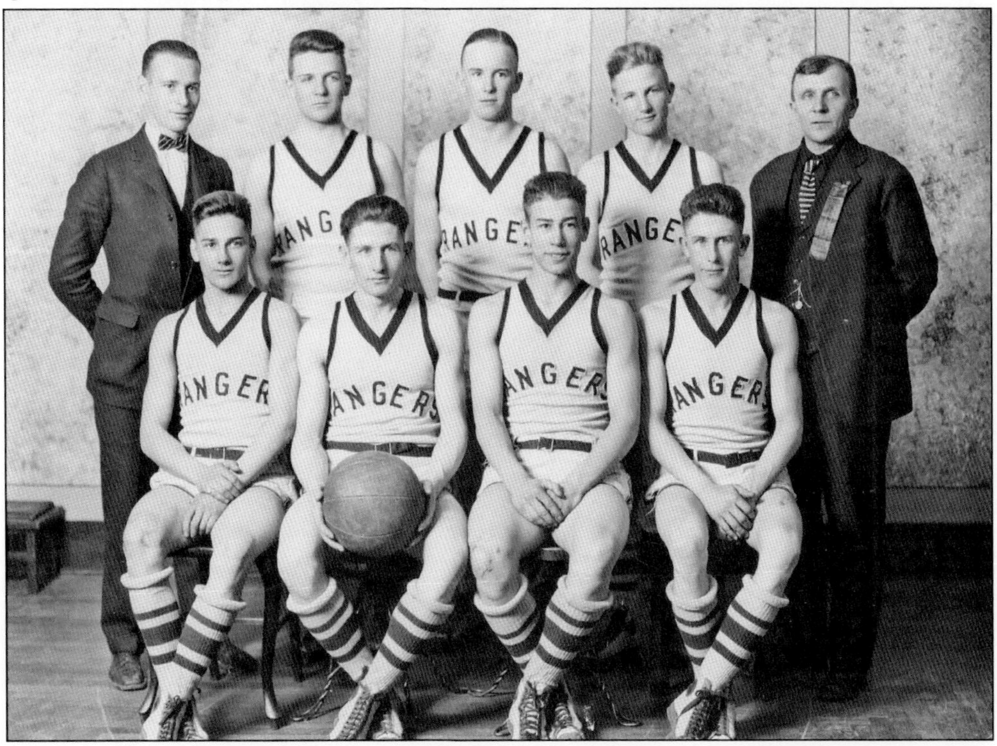

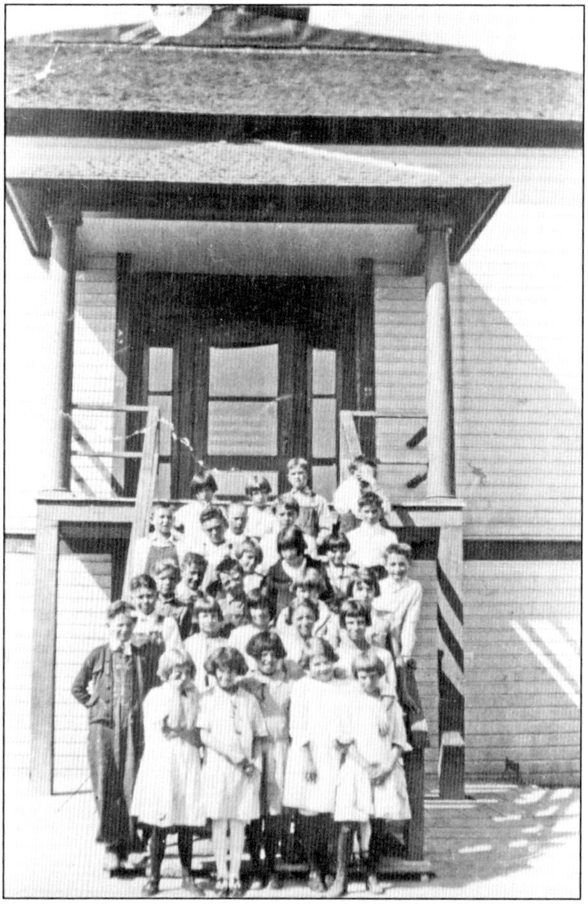

Most of the surrounding mining camps had a school. If the camp did not provide a schoolhouse, the classes were held in the company hall. The teachers would live in the town boardinghouse or with a reputable family in the community who could provide a room. The grades started with the first grade, and most ended with the ninth. All attended the same school, often in the same room. In one school in 1914, there were eight students with one teacher. By 1918, there were 98 students and two teachers. The students in the higher grades would have to come to Kemmerer to finish school. Most students graduated from Kemmerer High School. To the left is a class picture at the Frontier Grade School. The class photograph below is unidentified.

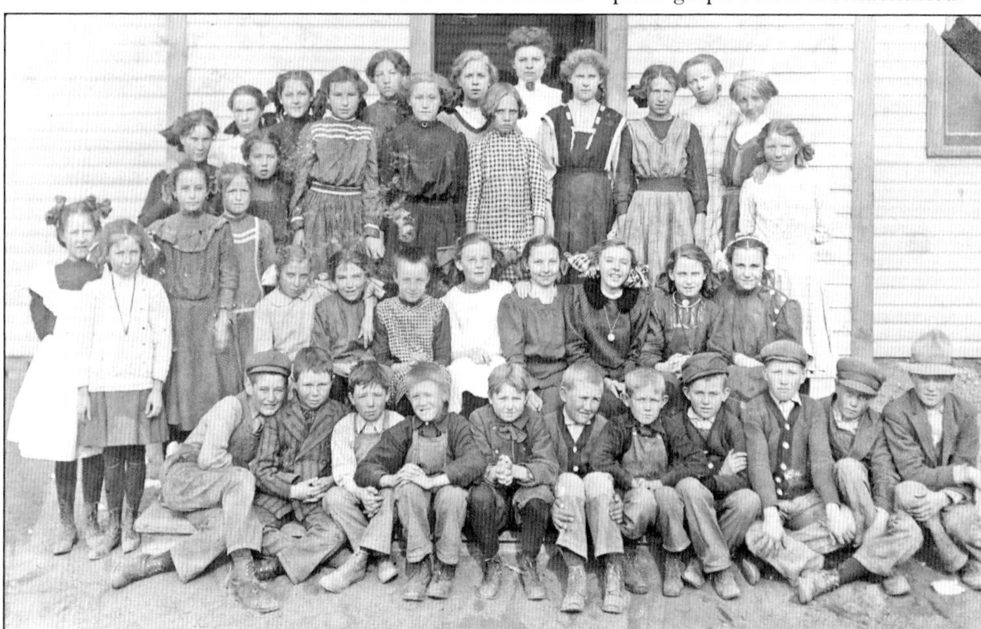

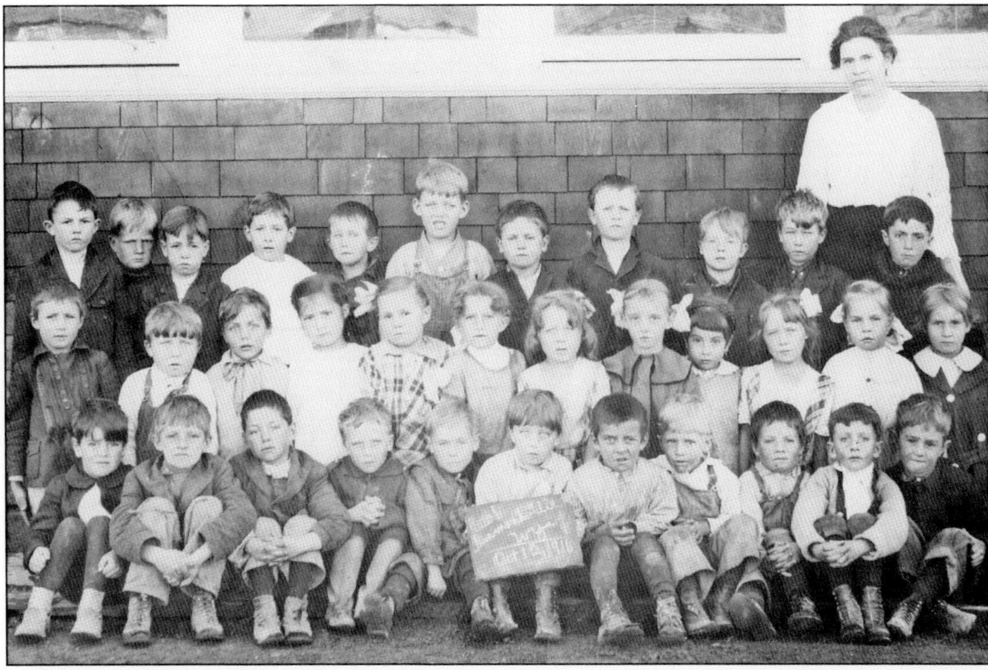

In 1897, Diamondville got its first school. School was originally held in a log house located on the east side of the Hams Fork River. In 1915, a new schoolhouse was built. The school was named after the superintendent, A. L. Burgoon. The photograph above shows the first-grade class in 1916; Mrs. Wright was the first-grade teacher. Burgoon, shown in the photograph below teaching a class in 1930, had taken the superintendent position in 1909. Under him, the Diamondville School was remodeled and made to accommodate the growing classes. Burgoon died December 25, 1949; he had been the superintendent of the Diamondville school district for 40 years.

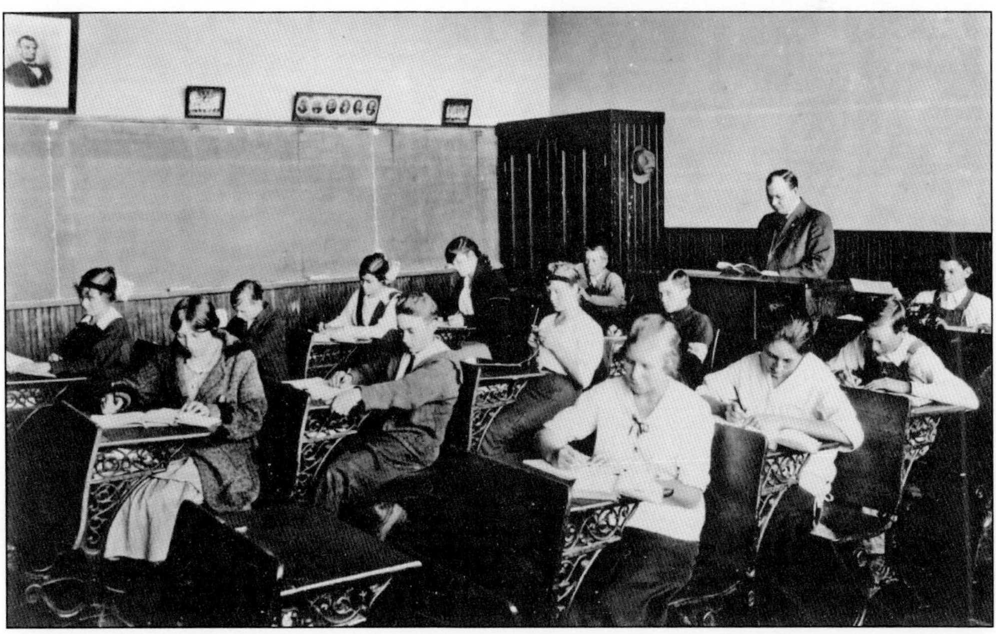

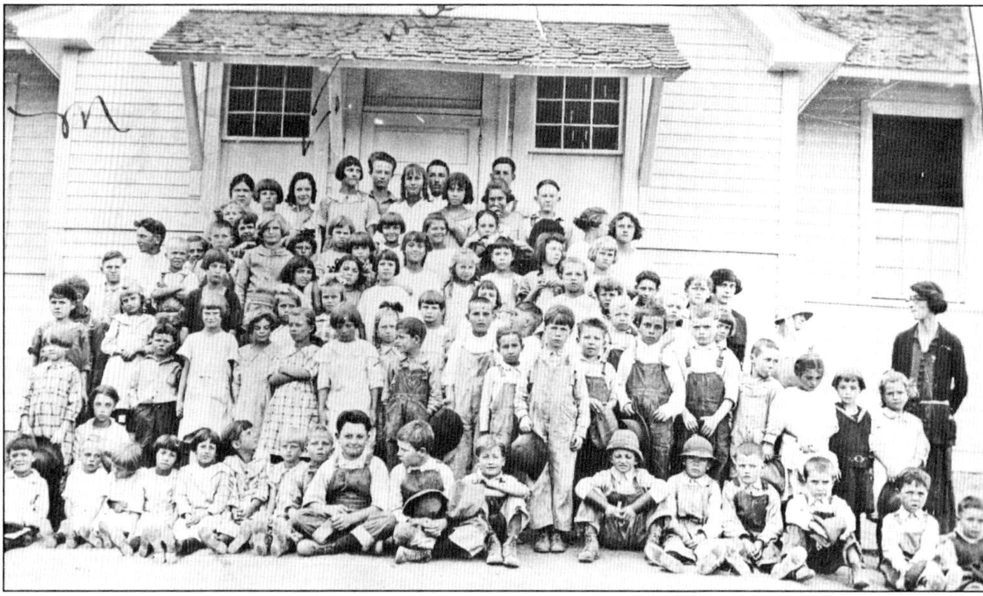

The above photograph shows students in 1928 or 1929. The students attended the Cumberland School in the coal mining camp of Cumberland. The school started in the 1920s. There was also a school located at the company ranch a few miles west of Cumberland. In 1930, when Cumberland closed down, all the children had to go to the company ranch school, which was funded by other school districts. The older students went to school in Kemmerer. The photograph below shows the graduating class of 1932. Some of the students were from the neighboring coal camps, but most lived in town. The ranches in the area all had their own teachers and schools for the younger students.

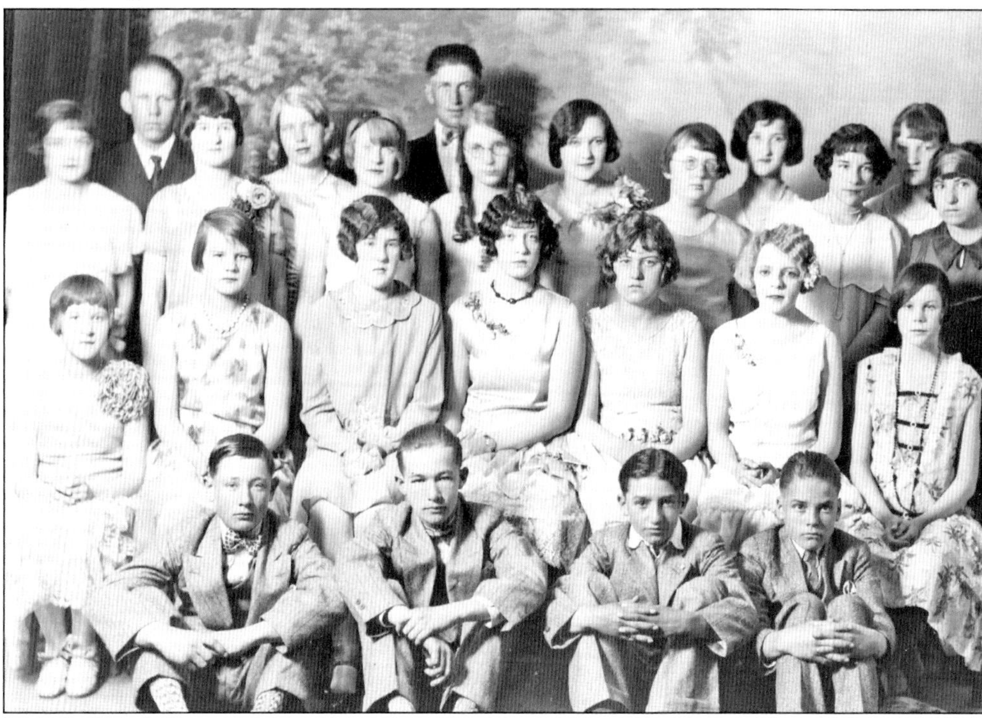

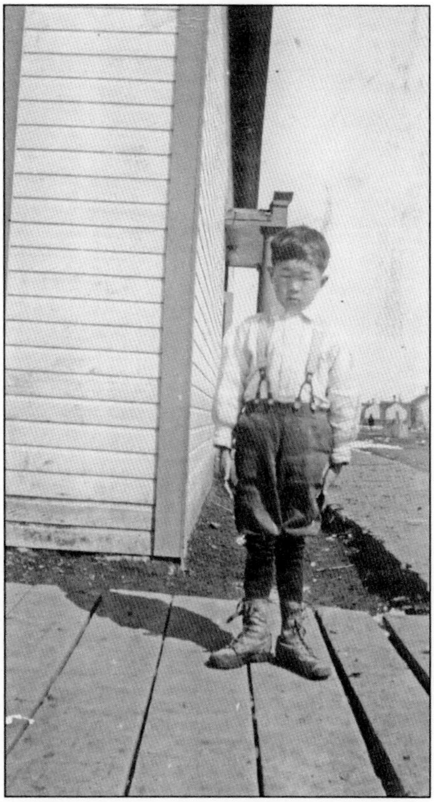

The town of Frontier had a large Japanese population; their houses were built on the present location of the golf course clubhouse. The Japanese had a bathhouse, pool hall, and a boardinghouse. But all of Frontier shared the same schoolhouse. When the children went to school, friendships formed with children from different ethnic groups. Everyone learned something about each of the different nationalities that lived in Frontier. The adults kept mostly to themselves, but the children did not. After school, most of the children would gather to play games, go fishing, or, in the winter, sledding and ice skating.

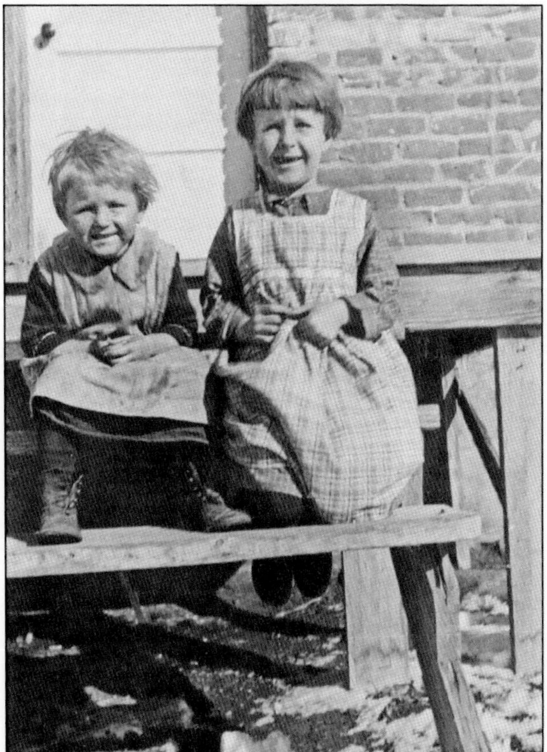

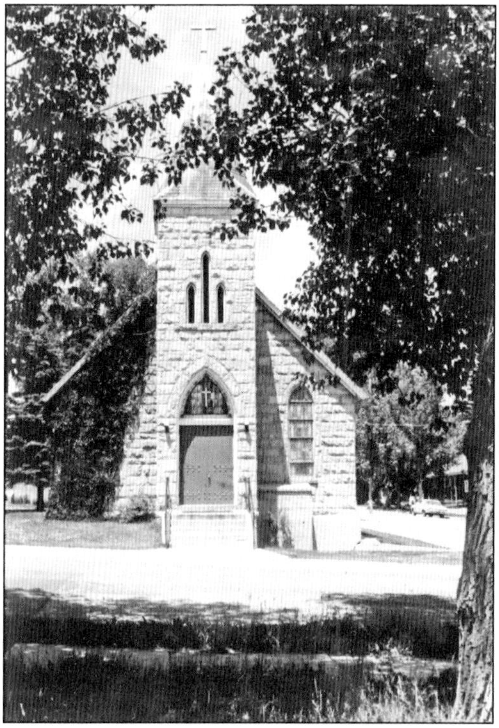

Kemmerer's first Catholic church was erected on a lot donated in 1899 by the Kemmerer Coal Company. The first work on this stone church began the following year by the Reverend Peter Casey. Reverend Casey also gave Kemmerer's first mass in 1898 in Keenan's Hall. The cost of this new edifice was $7,500, half of which was collected from the parishioners and the other half borrowed. Fr. M. Golden was the first resident priest of this newly formed St. Patrick's Parish in 1901. The choir loft was built in 1919 by the Reverend Gerard Schellinger.

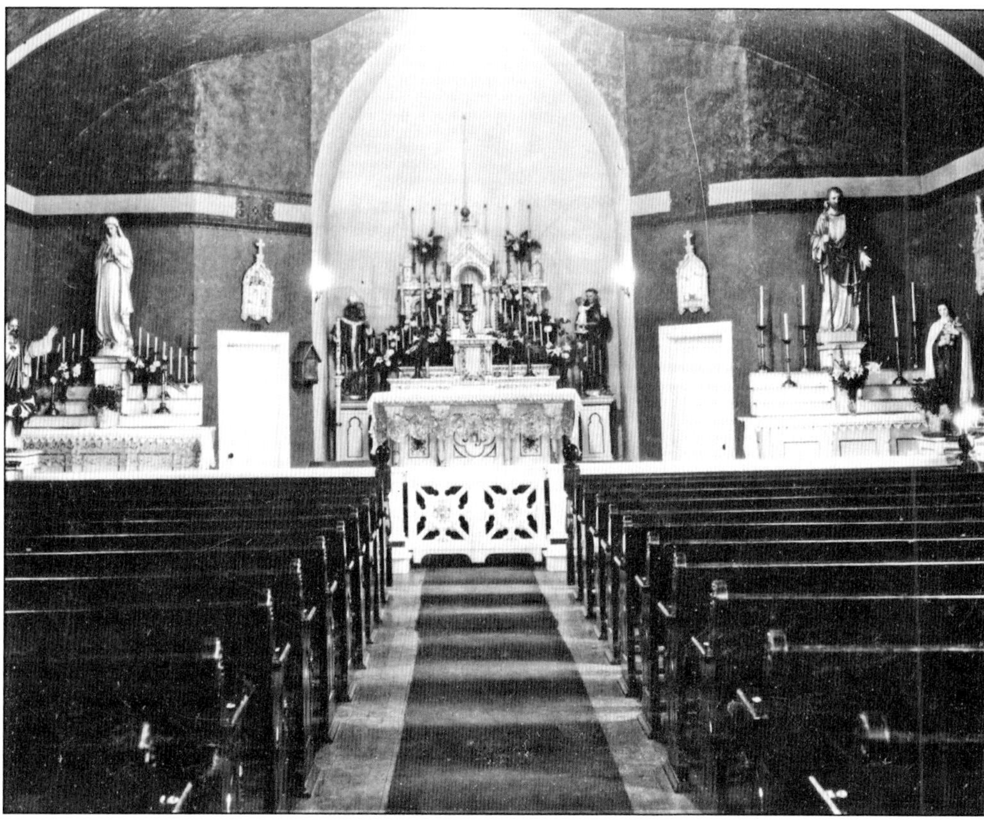

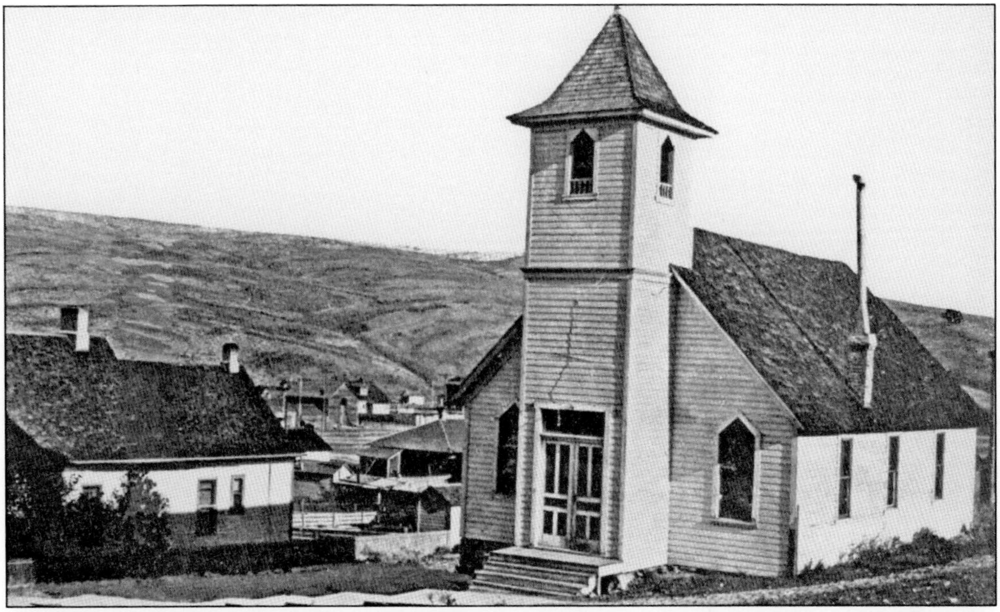

Kemmerer's first Methodist Episcopal church was completed in November 1899 and built on land donated by the Oregon Short Line Land and Development Company. The church was dedicated on December 1, 1899. The first minister to serve in this church was the Reverend J. Putman, serving from May 1899 to January 1900. This small building served Kemmerer's Methodist community for 49 years. A larger Methodist church building was constructed about 1950 with a large portion of the money coming from J. C. Penney; therefore, the church was named the Berta A. Penney Memorial Methodist Church. The Episcopal church also built a larger building in 1920. The church was named the St. James Episcopal Church. The bells in the church tower were given as a memorial by Pete and Elna Jiacoletti and are referred to as the Jiacoletti Bells.

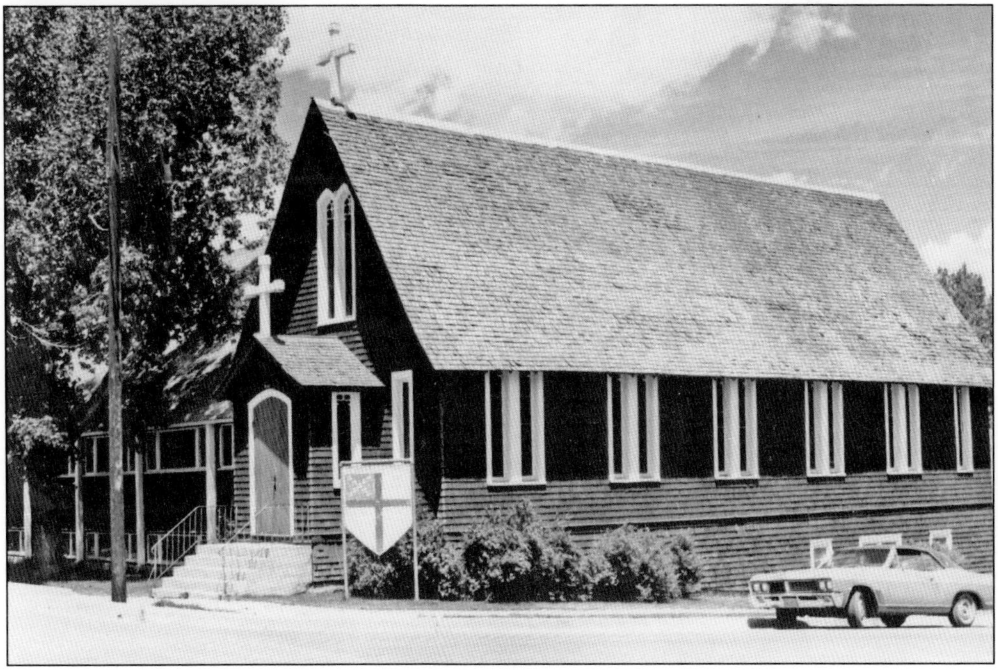

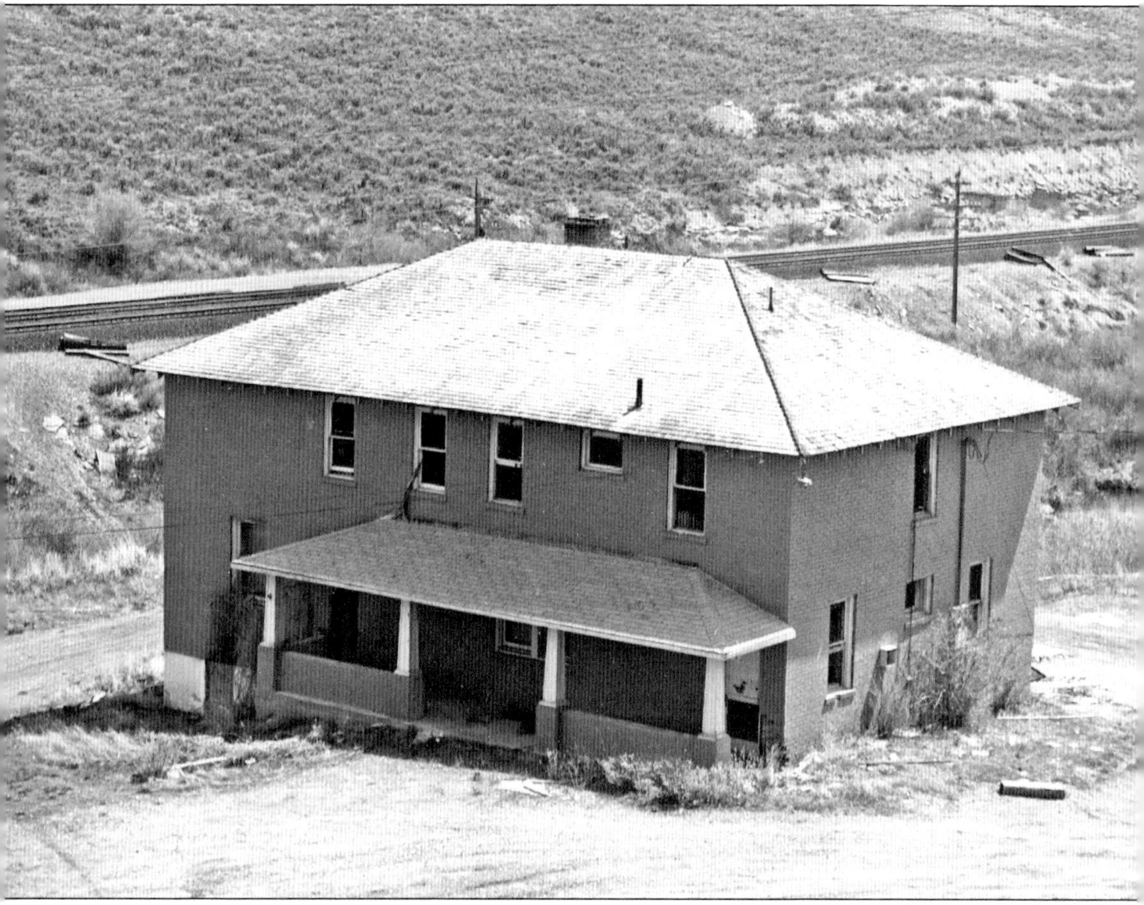

Having enjoyed a truly legendary and colorful history, Kemmerer's Southern Hotel operated for seven decades as a full-fledged house of ill repute. In 1898, Isabelle Burns, a "sporting lady" from Salt Lake City, boarded a train heading east looking for greener pastures. She had a small inheritance and her goal was to open a house of ill repute. In Kemmerer, Madam Isabelle found a dance hall with adjoining rooms that was in financial straits. She cleared the debt and was in business. Her new business was called the Green House; this building was eventually moved to a location next to the railroad and renamed the Southern Hotel. The house operated for the next 30 years. In June 1926, the house burned to the ground. The fire resulted in the death of one of the ladies. A new house was built and operated for the next 40 years as a brothel, which was common knowledge in the community. On January 17, 1967, Lincoln County's newly elected sheriff closed the Southern Hotel because it was in violation of Wyoming's law on "houses of ill fame."

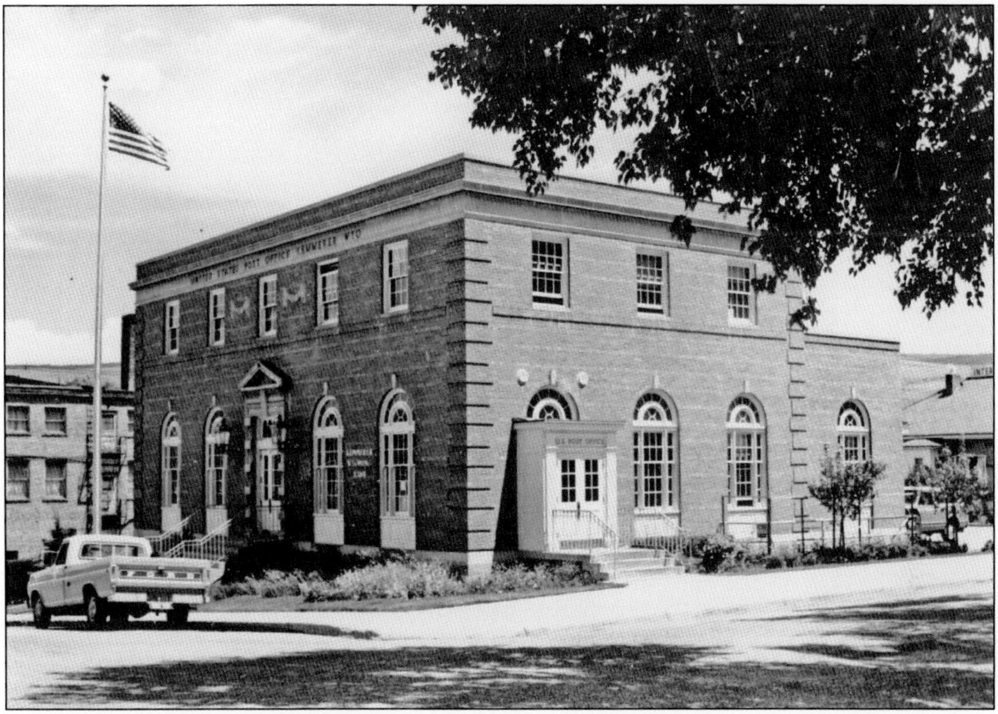

Kemmerer's third and final post office was built in late 1935. The building was also used to house the national forest office. The post office is located on the corner of Sapphire and Cedar Streets. On February 15, 1936, a formal opening of the building took place with state senator Platt Wilson as master of ceremonies. On the wall inside the post office are two hand-painted murals. They depict a triceratops along with other fossilized remains of dinosaurs. These murals represent all the fossils that have been found in the area, making Kemmerer the "Fossil Fish Capital of the World." The murals were painted by Eugene Kingman in 1938.

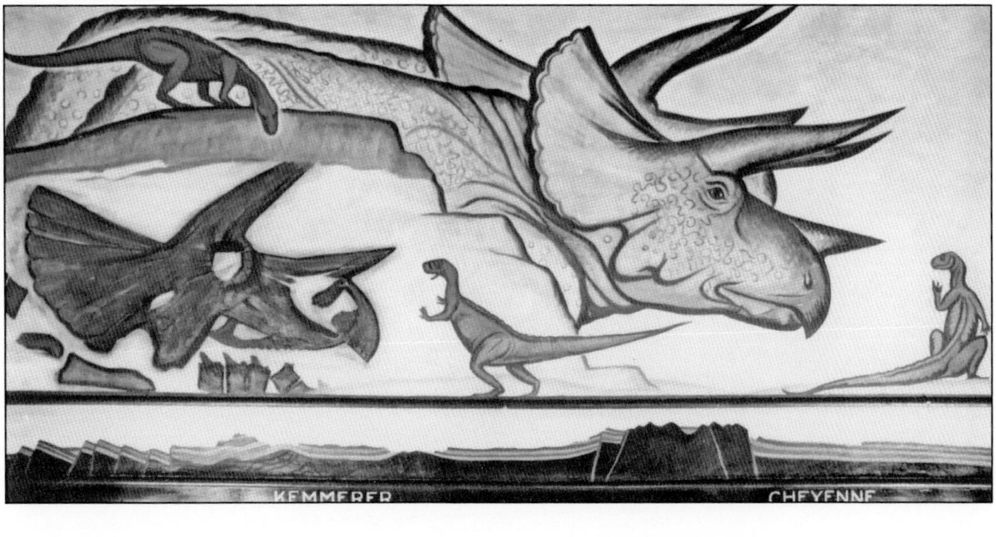

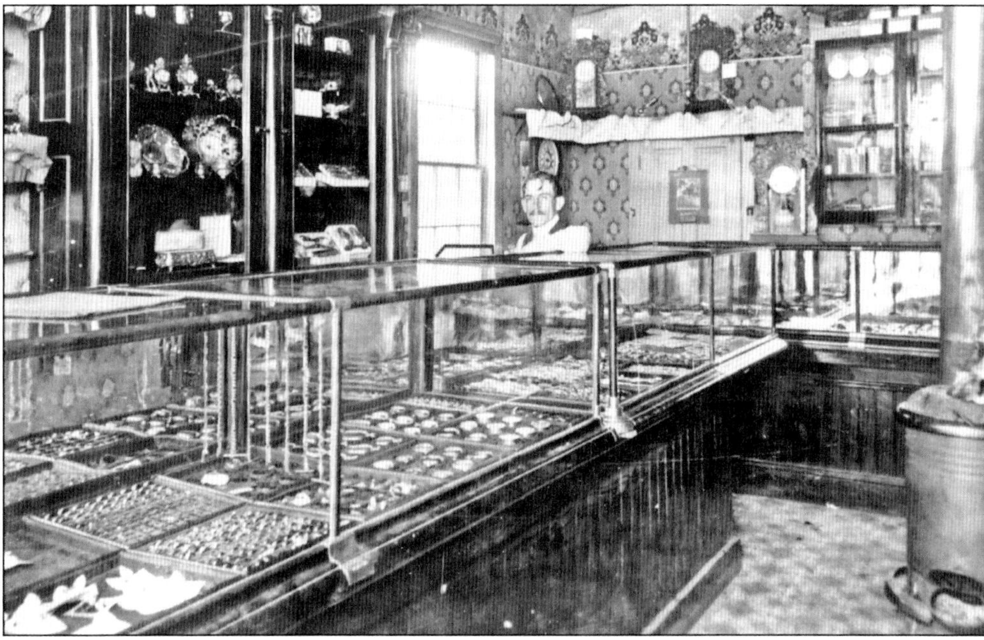

One of the early stores on the Triangle in Kemmerer was the F. E. Curtis Jewelry Store. Curtis opened the store in Kemmerer in 1906. The store was originally located on North Main Street; in 1916, it was moved to a location on Pine Avenue. The building was demolished in 1979 so the parking lot of the bank could be enlarged. The store was closed after being in business for 73 years. Curtis ran his business until he was 93 years old. Because he smoked a pipe or a cigar, the store items would smell like tobacco. When Curtis died in 1981, he was believed to be the oldest resident in Kemmerer. The exterior of the jewelry store is shown below, far left.

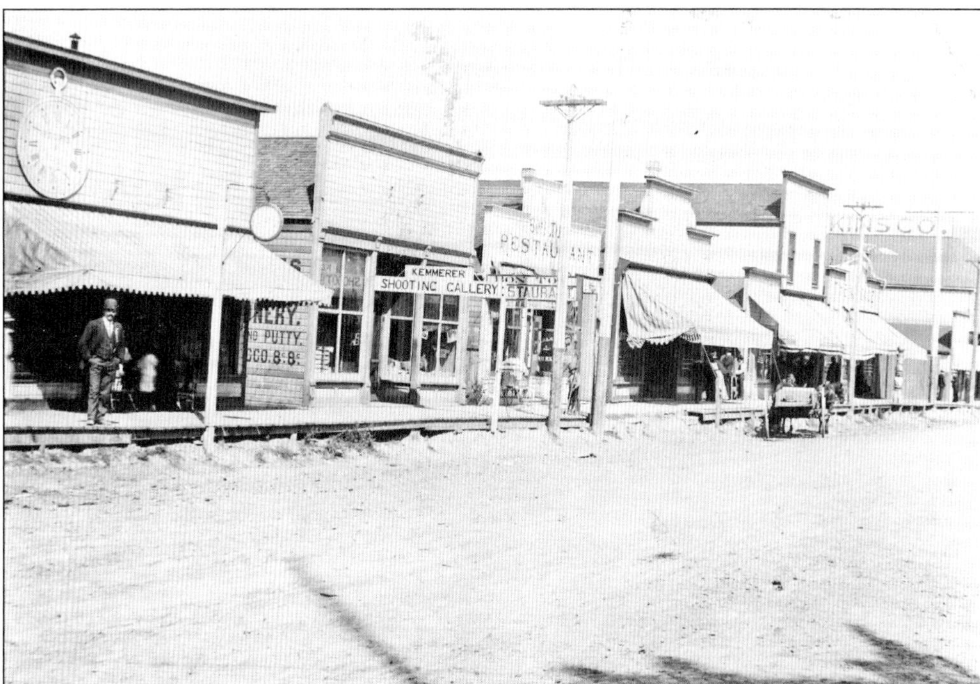

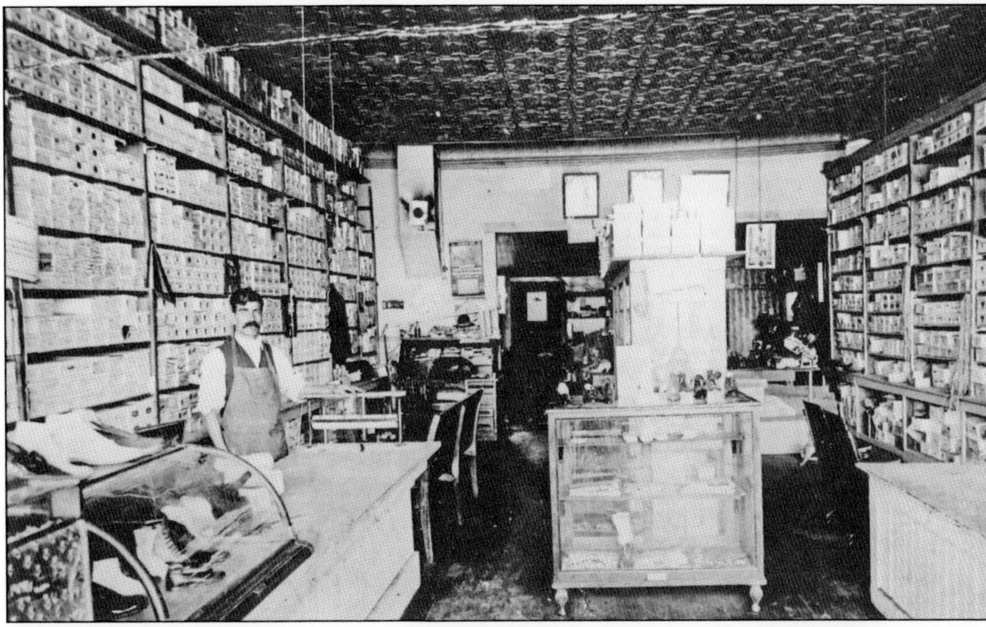

The oldest family-owned business in Kemmerer is Sawaya's Shoe Store. It was opened in 1909 by Sabeh Sawaya (above). He emigrated from Lebanon in 1900, first settling in Colorado before coming to Kemmerer. Sawaya not only sold shoes but made and repaired them; he learned this trade in Lebanon. The shoes he made were of fine leather and fastened with buttons. In 1927, Sabeh retired and turned the business over to his oldest son, George (below, right). George expanded the store, adding an upstairs. In 1972, George retired and turned the business over to his son John (below, left) and John's wife, Mary. John and Mary Sawaya continue to run the store today.

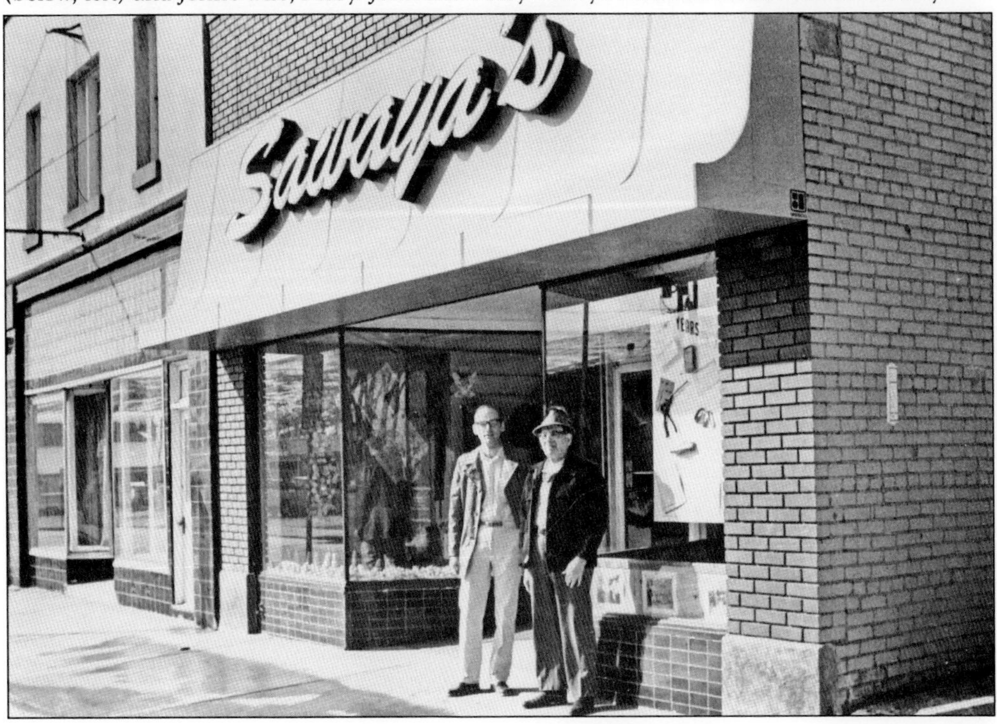

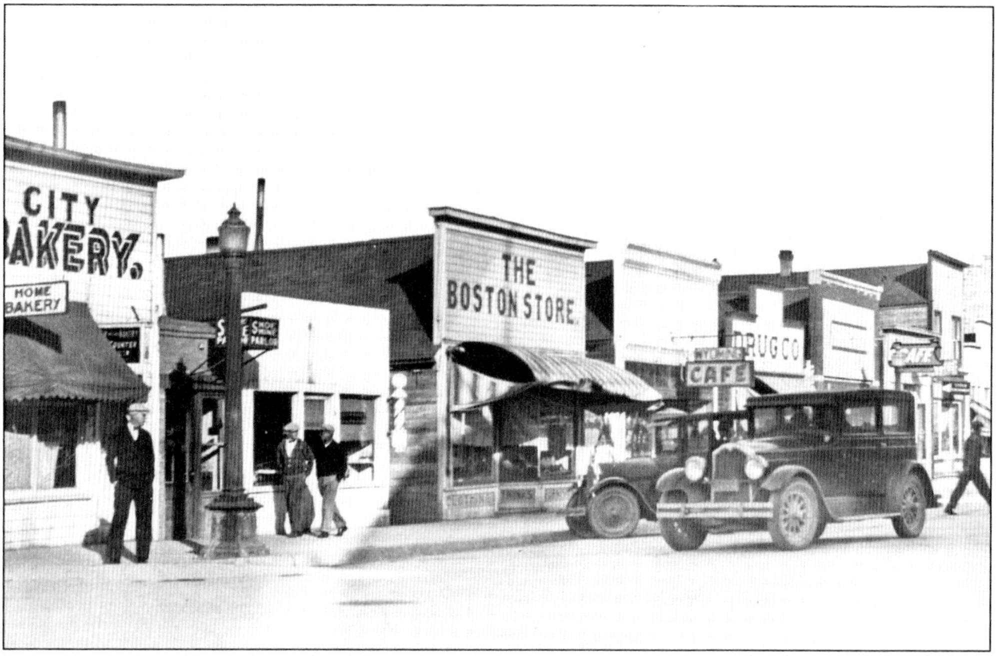

Various businesses started in Kemmerer; some were successful staying in business for many years and others were not so successful, only lasting for a short time. During the 1920s, the business district changed dramatically. According to the State Business Directory of Wyoming, during the 1920s, Kemmerer had 13 saloons, 9 general merchandise stores, and a variety of other businesses and hotels. When Prohibition was passed, the saloons around the Triangle had to change what they offered. Some of the buildings shown in these photographs are still standing and being used. In 1934, more progress was made to the downtown area when the town council decided to add more sidewalks and oil the main street through Kemmerer.

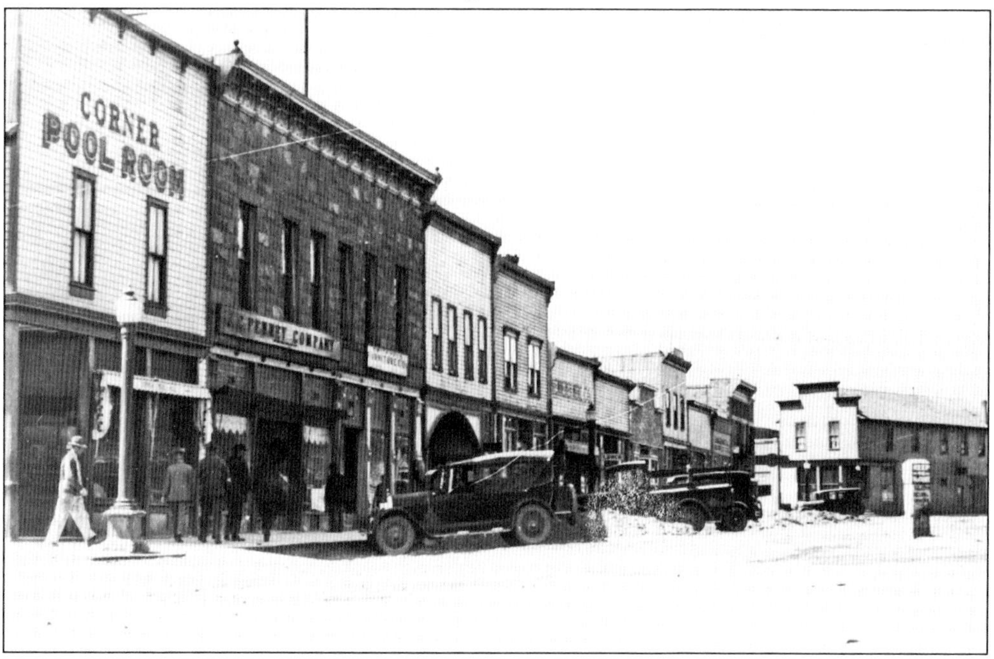

Six

People, Events, and Places

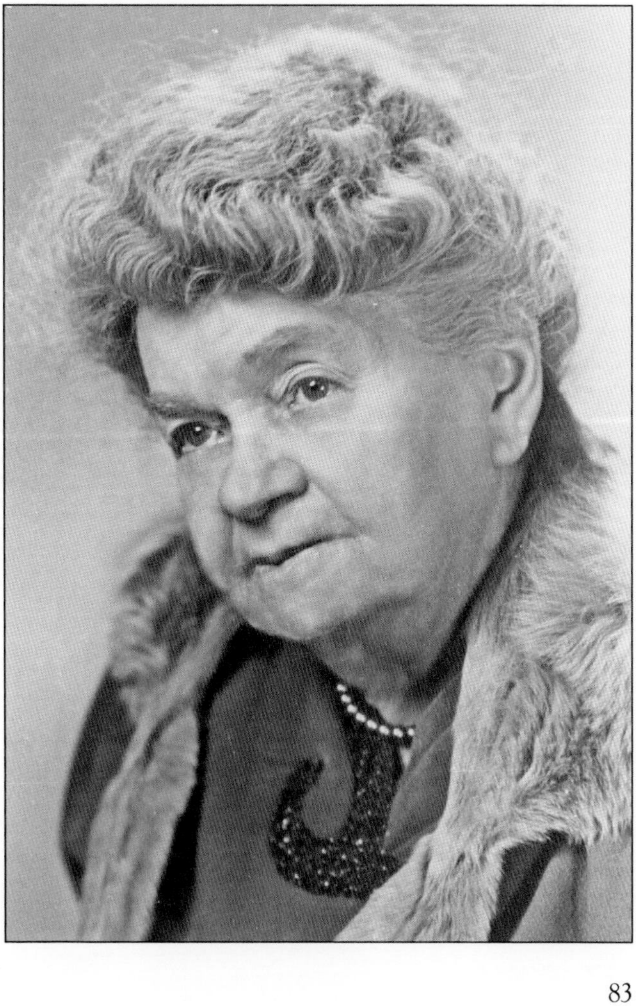

One of Kemmerer's most influential ladies was Susan Jane Quealy. She married Patrick Quealy in 1890 and moved to the Kemmerer area in 1900. She was active in planning community activities, especially for children. She was also active in various women's groups. Her selfless service and kindness were felt by the whole community.

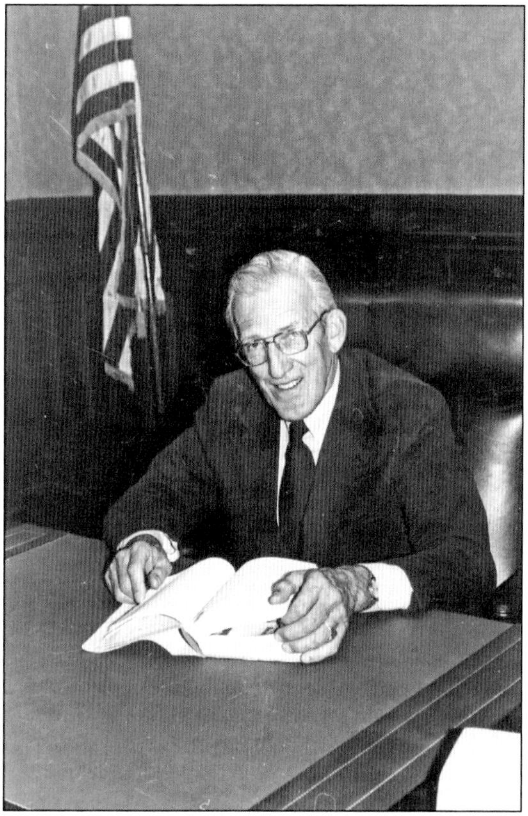

Edgar Jacob Herschler was the 28th governor of the state of Wyoming. He was the only three-term governor in Wyoming history. He received his law degree from the University of Wyoming in 1949. He served in the Wyoming House of Representatives for 10 years, from 1959 to 1969. He was the city attorney for Kemmerer as well as the prosecutor for Lincoln County. He served as governor from 1975 until January 1987. He is best known for his slogan "growth on our terms," which referred to the energy boom in Wyoming. Herschler was born October 7, 1918, on Fontenelle Creek, just north of Kemmerer. He died February 5, 1990, in Cheyenne, Wyoming. A sign was erected on the Triangle Park in June 1990 renaming the park Herschler Triangle Park, honoring Kemmerer's native son and his accomplishments.

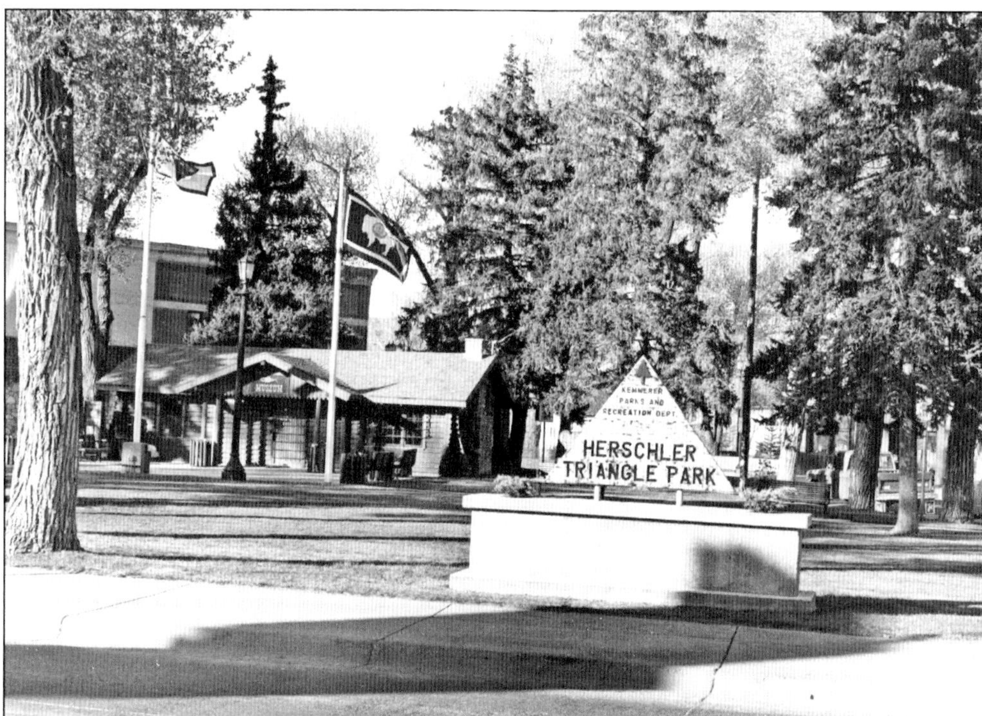

Two additional monuments stand in Herschler Triangle Park. The first is dedicated to the hardworking men and women who first came to the area. They were able to tame and settle in this part of southwest Wyoming and call it home. They helped to make the area what it is today.

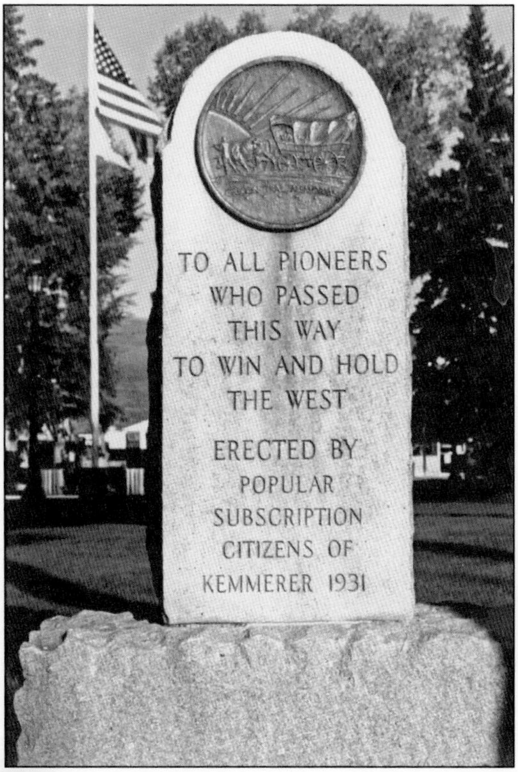

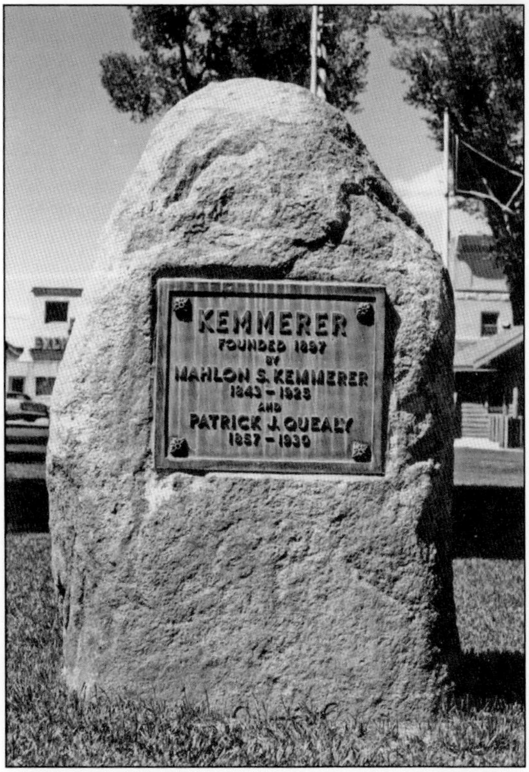

The second monument that stands in the park is dedicated to Kemmerer's founding fathers, Mahlon S. Kemmerer and Patrick J. Quealy. These two men had the foresight and knowledge to be able to create the town of Kemmerer and give her the push needed for continued growth. Kemmerer is the place it is today because of these men.

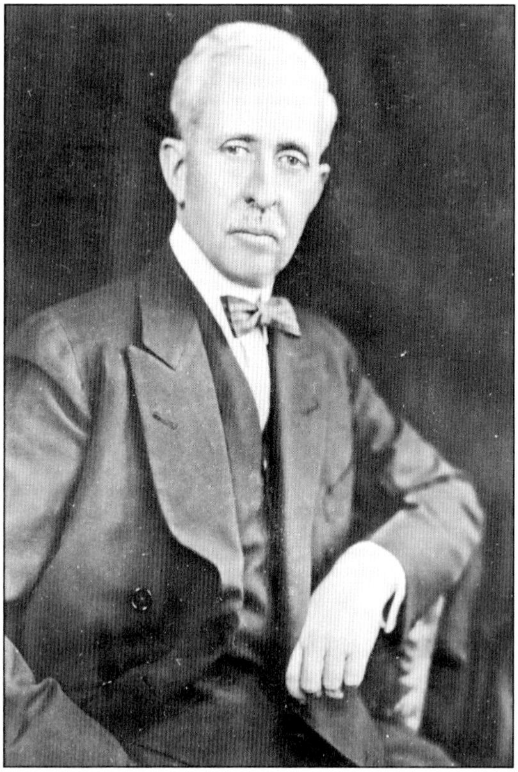

One of the most famous people to live in Kemmerer was retail giant James Cash Penney (pictured here). He was born in Hamilton, Missouri, on September 16, 1875. His goal was to live to be 100 years old. Although he did not achieve this goal, dying at the age of 95 on February 12, 1971, he did accomplish great success: he was able to build an empire. His business career began at age eight when he had to earn money to buy his clothes. After high school, he was trained as a merchant. In 1898, he moved to Colorado and started working for Guy Johnson and Thomas Callahan.

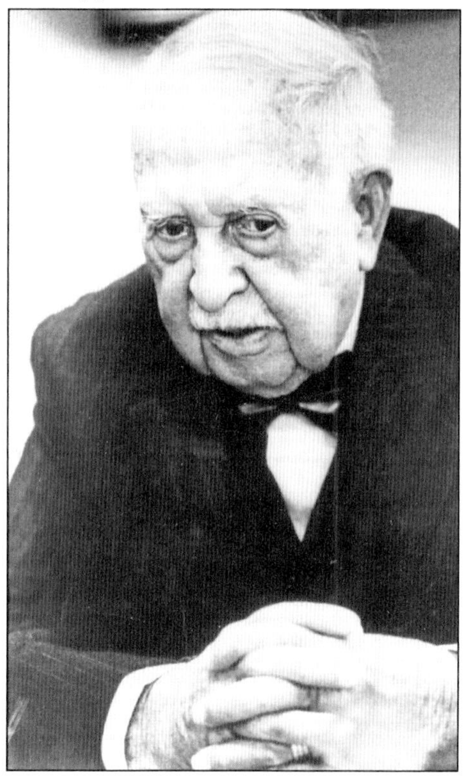

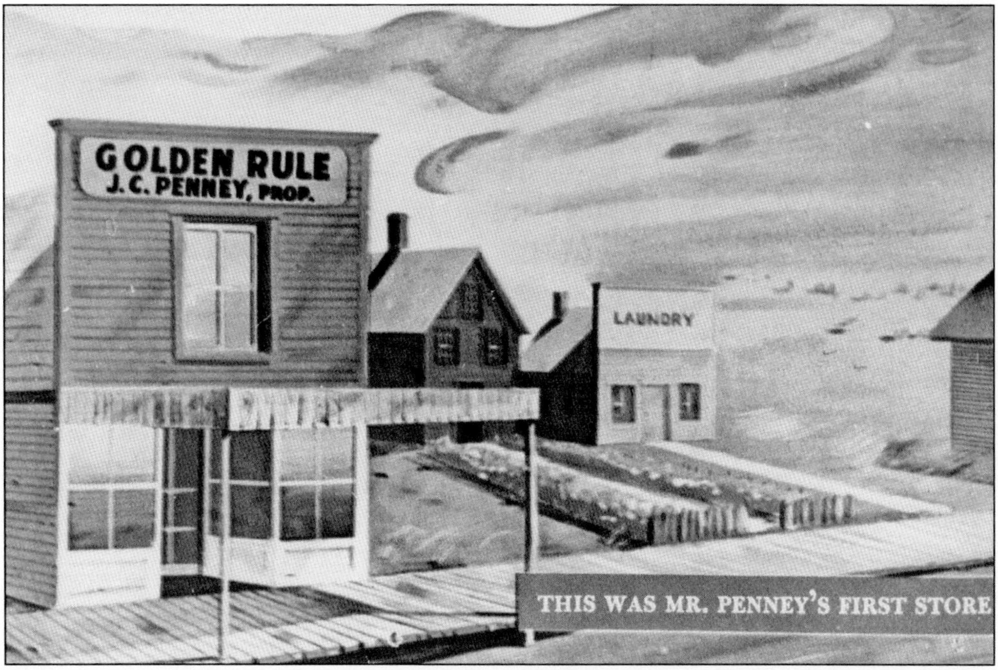

Johnson and Callahan operated a dry goods store called the Golden Rule in Colorado and Wyoming. Penney was able to buy into the business and moved his family to Kemmerer. On April 14, 1902, Penney opened the Golden Rule store in Kemmerer on Pine Avenue. Penney was told that he was doomed to fail, but the community supported him and the store, which was a success on the first day of business. Penney wanted his store to be cash only, and in a mining community, this was not the norm, as the miners charged at the company store. But the business continued to grow, and in August 1904, the store was moved to a larger stone building on South Main Street.

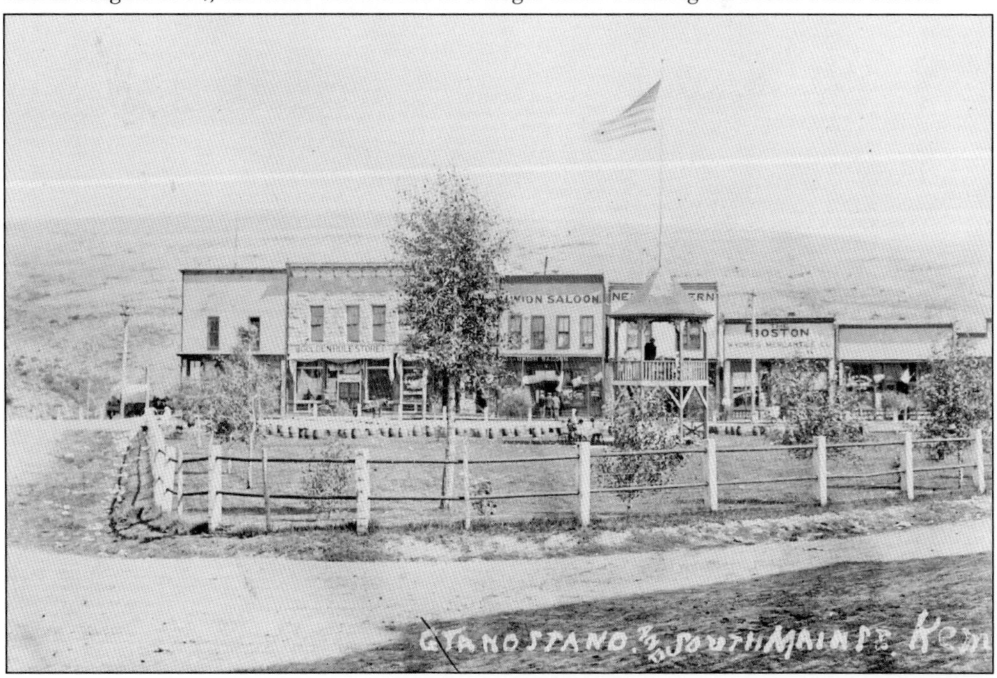

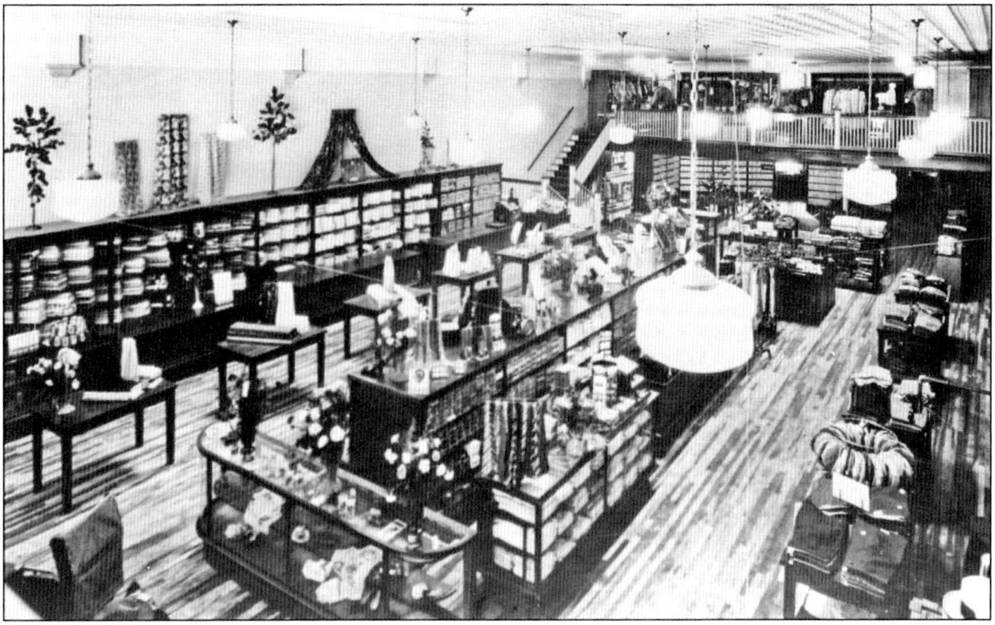

In 1907, Penney bought out his partners, Johnson and Callahan, becoming sole owner of the Golden Rule stores. The store kept the name until 1913, when it was incorporated as the J. C. Penney Company. The Kemmerer store is the first store, or "mother store," of the company. By 1929, the store had outgrown its location and was moved for a third and final time to the present location on the corner of Pine Avenue and J. C. Penney Drive (pictured here). The Penney home was moved to a location on J. C. Penney Drive and is open as a J. C. Penney museum. Penney stated at a Kemmerer businessmen's meeting that he would always have a warm spot in his heart for Kemmerer.

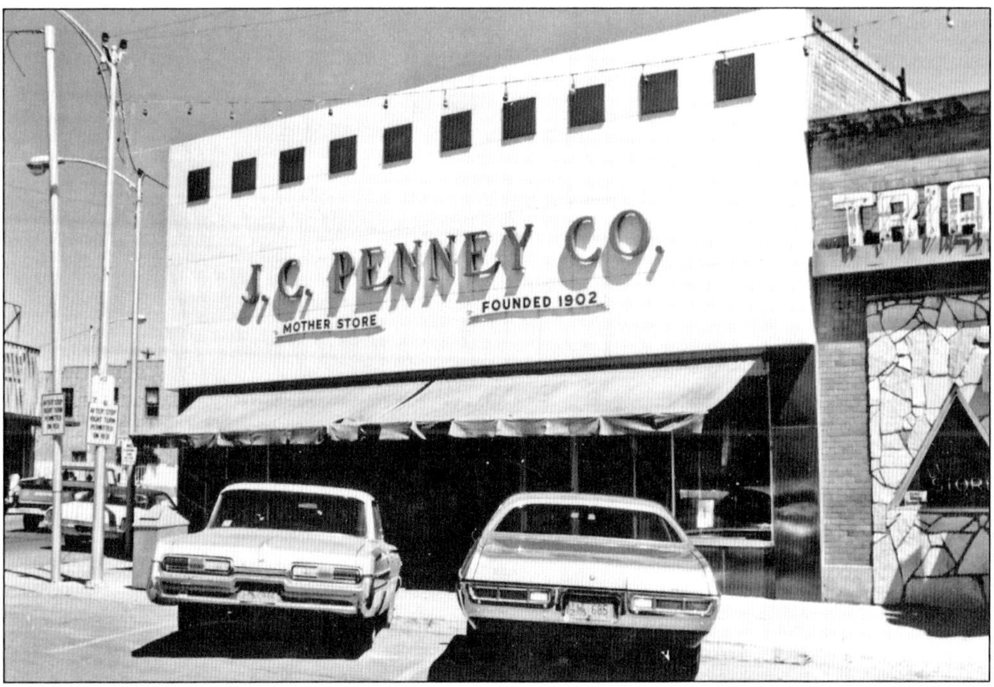

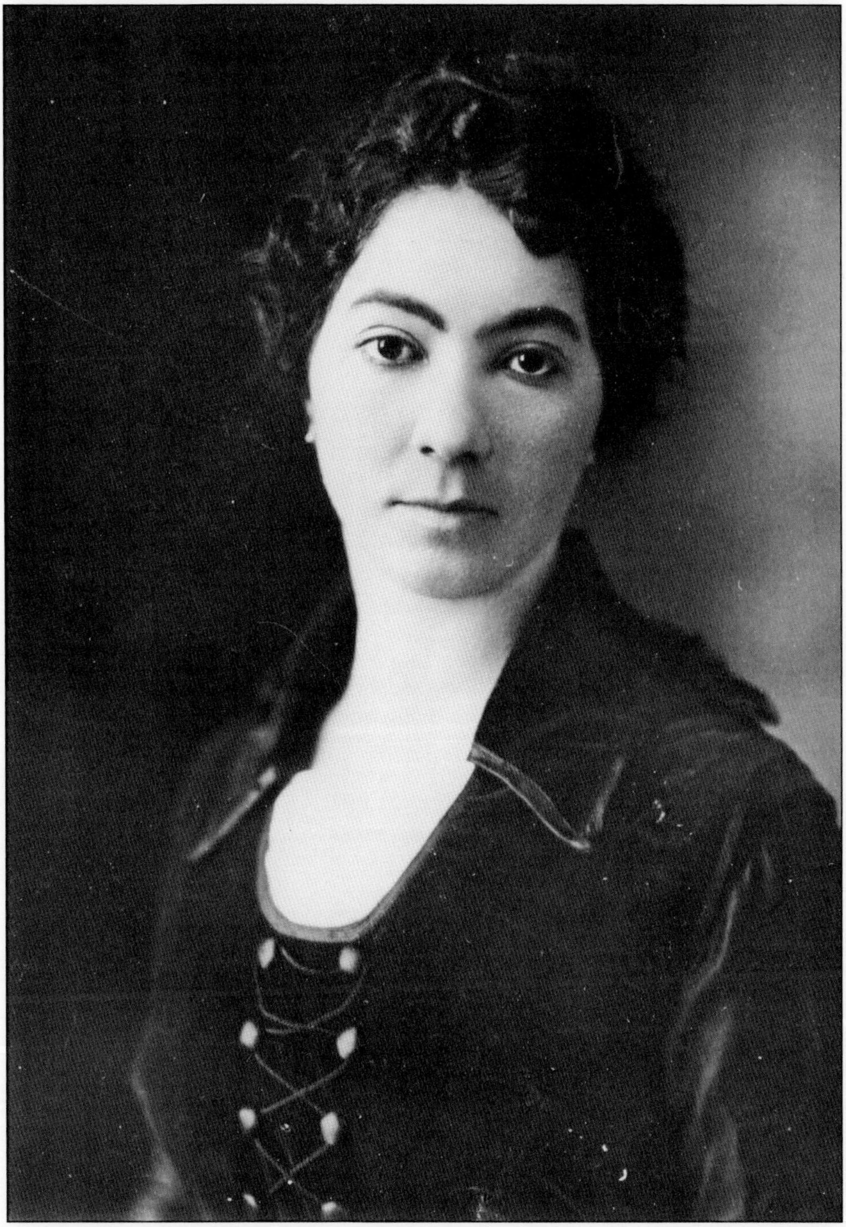

Kemmerer has its own unsolved mystery, the death of cattle rustler Annie Richey. Anna Byers Richey, also known as Annie, was born May 19, 1890, in Fossil, Wyoming. She was known in the area as the Petticoat Rustler. She is the only woman in Wyoming ever to be convicted of cattle rustling. She was married in the fall of 1907 to a local schoolteacher, Curtis Richey. The newlyweds were given the Byers family ranch on Rock Creek; the couple divorced after 11 years of marriage. Annie stayed on the ranch and continued to raise cattle. During the fall roundups and shipping, it was alleged that Annie would gather and ship more cattle than belonged to her. In August 1919, she was arrested for cattle rustling. On November 25, 1919, she was found guilty and sentenced to six years in prison at the Colorado State Penitentiary. Annie was released on bail so she could take care of personal business before going to prison. But before she went to prison, someone poisoned her, and she died on May 19, 1920.

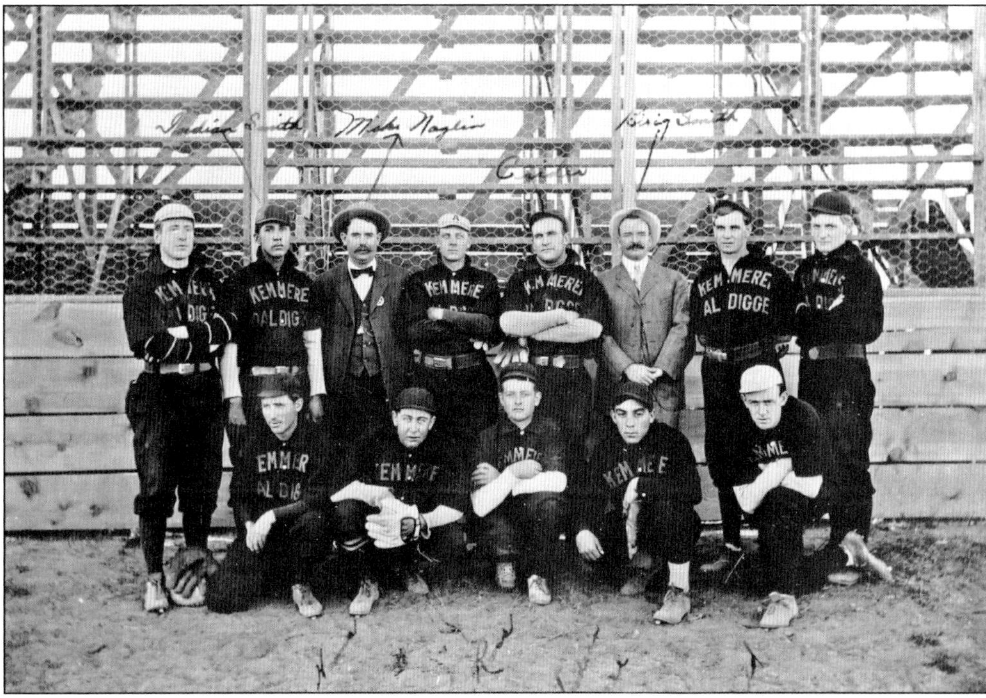

Baseball has been a popular sport in the Kemmerer area since the coal mining camps began in the 1890s. All the coal camps had a baseball team; they would practice in the sagebrush but would have to come to Kemmerer, Frontier, or Diamondville to play a game. Diamondville's field had a large covered grandstand, and Kemmerer's field was located where the Lincoln County Courthouse now stands. This field was also used for football games. The 1908 Kemmerer Coal Company team was called the Kemmerer Coal Diggers, and Quealy provided the team uniforms. The rivalry between Kemmerer and Diamondville was for the most part intense. Players were employed by the two local companies; men who were skilled baseball players were brought in and provided the jobs they wanted in order to strengthen the teams.

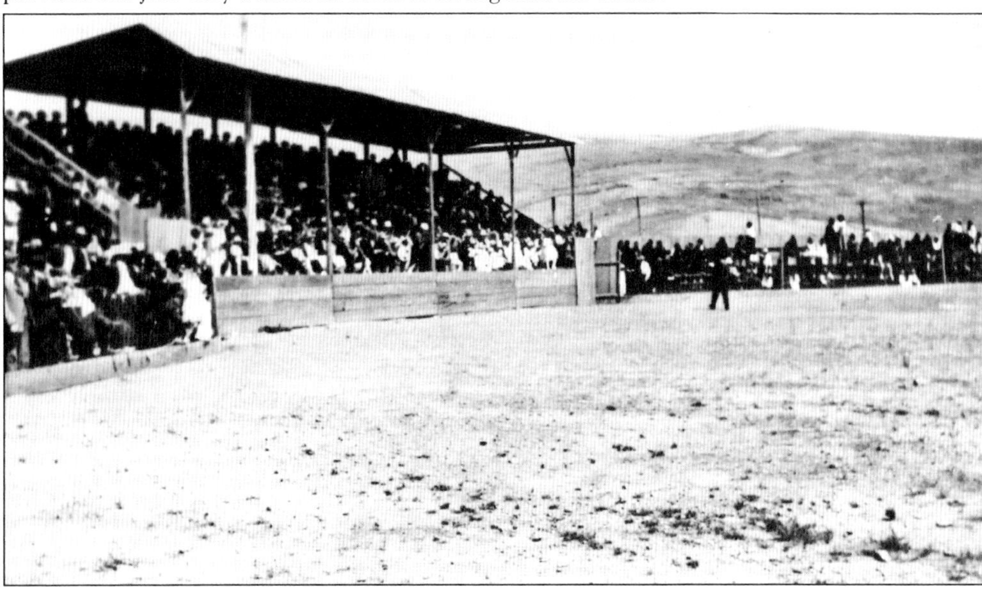

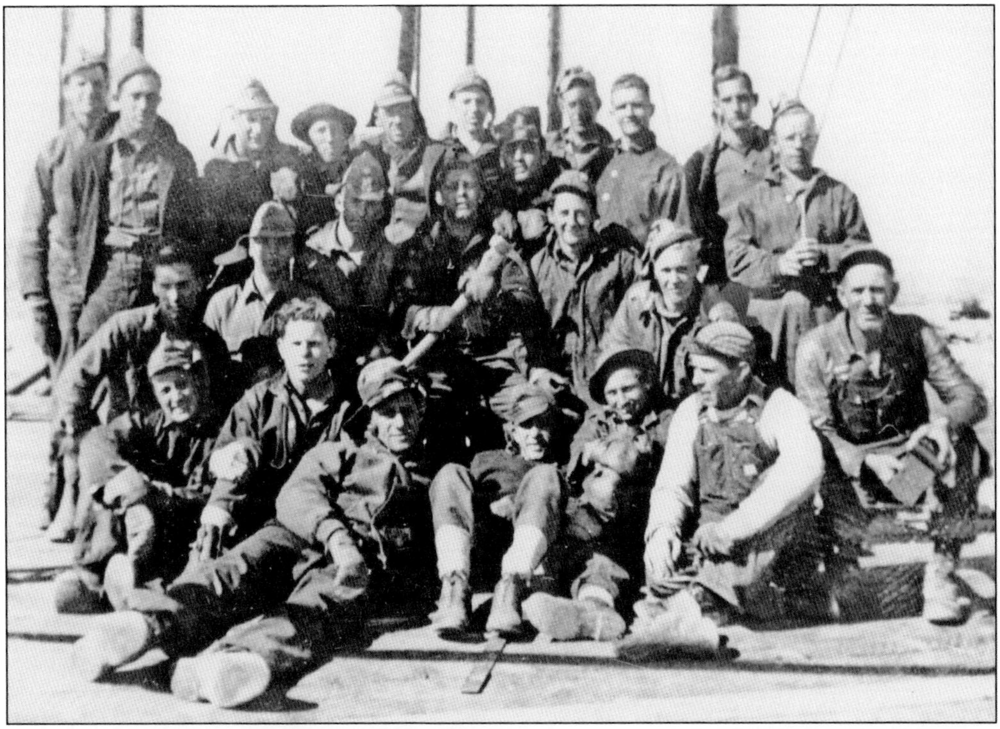

Nearly 3 million young men from 1933 to 1942 made up the Civilian Conservation Corps (CCC). They lived in camps like the one pictured. The camp in Kemmerer was located on a flat next to the river between Kemmerer and Frontier. The work was physically demanding as they built roads and reservoirs and improved the infrastructures of towns, forests, and historical sites. In 1933, during the Depression, Pres. Franklin D. Roosevelt passed the New Deal program; the goal was to put America back to work. One of its programs was the CCC. The corps disbanded in 1942 when many of the young men went to fight in World War II.

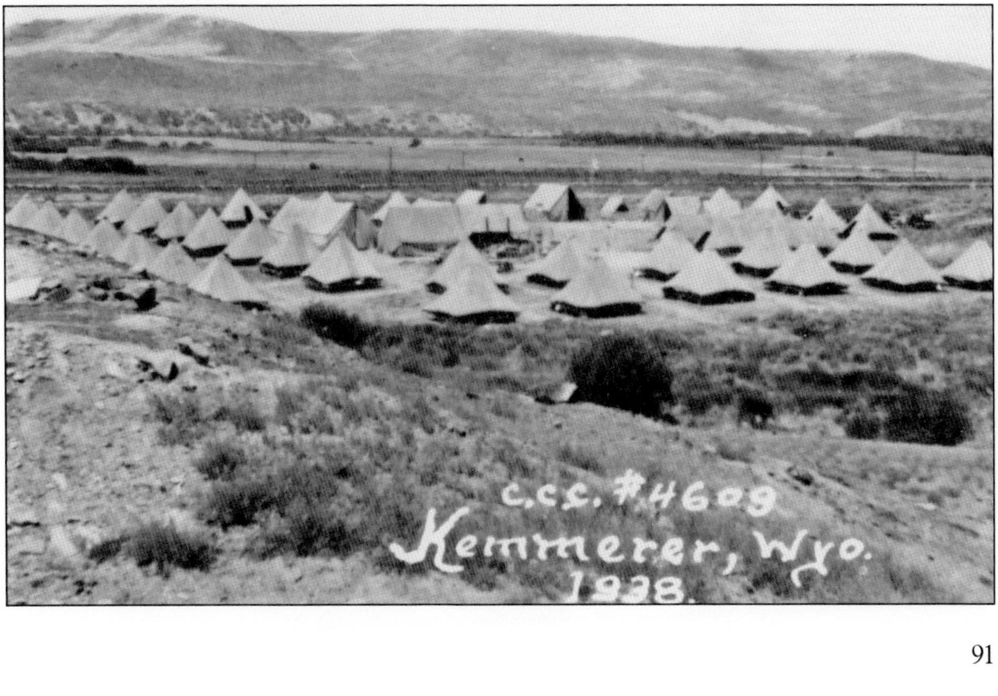

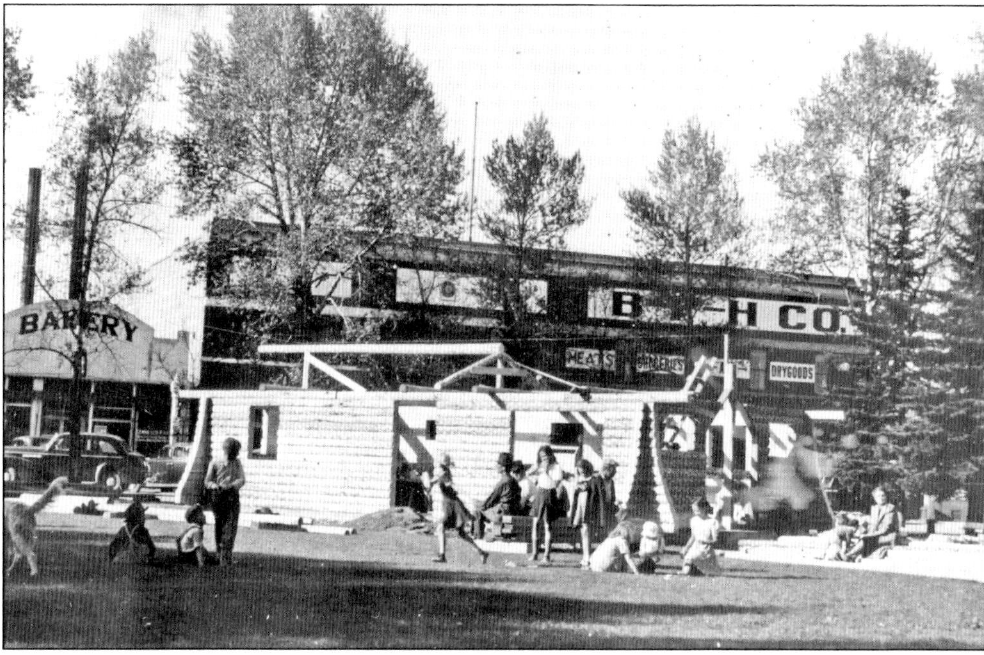

In early 1940, the town council decided that Kemmerer needed a museum. The construction of the building began in late 1940. The building is located at the top of the triangle in the center of town. The building was used as a museum for many years and was run by a core of dedicated volunteers during the summer months. The museum artifacts eventually outgrew the building, and a larger place had to be found. The building is now the home of the Kemmerer/Diamondville Chamber of Commerce. It serves as a meeting place for the summer activities held in the park.

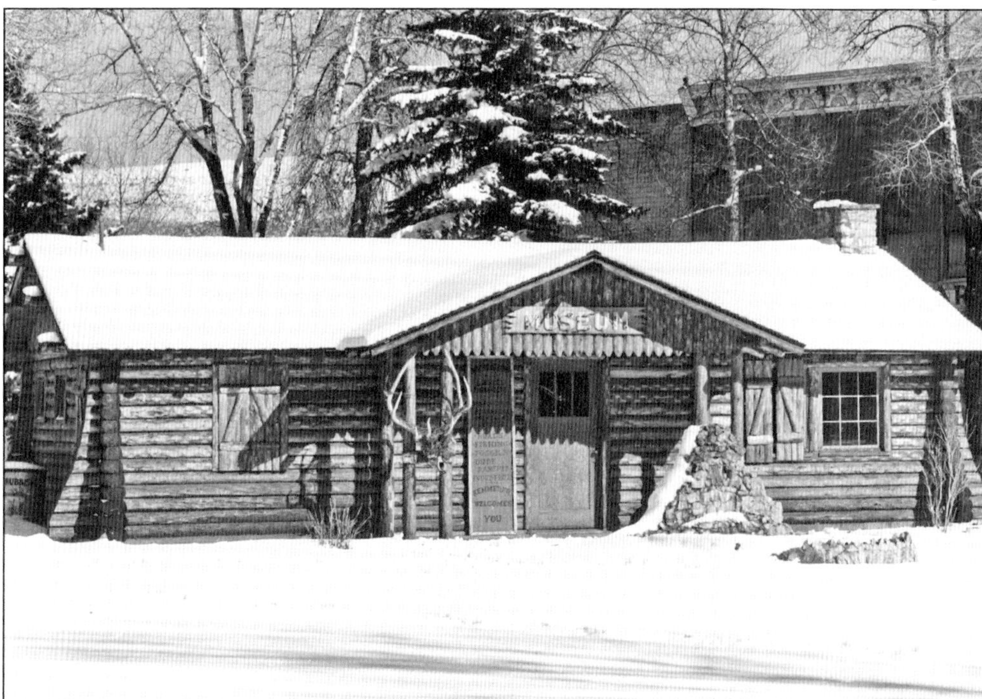

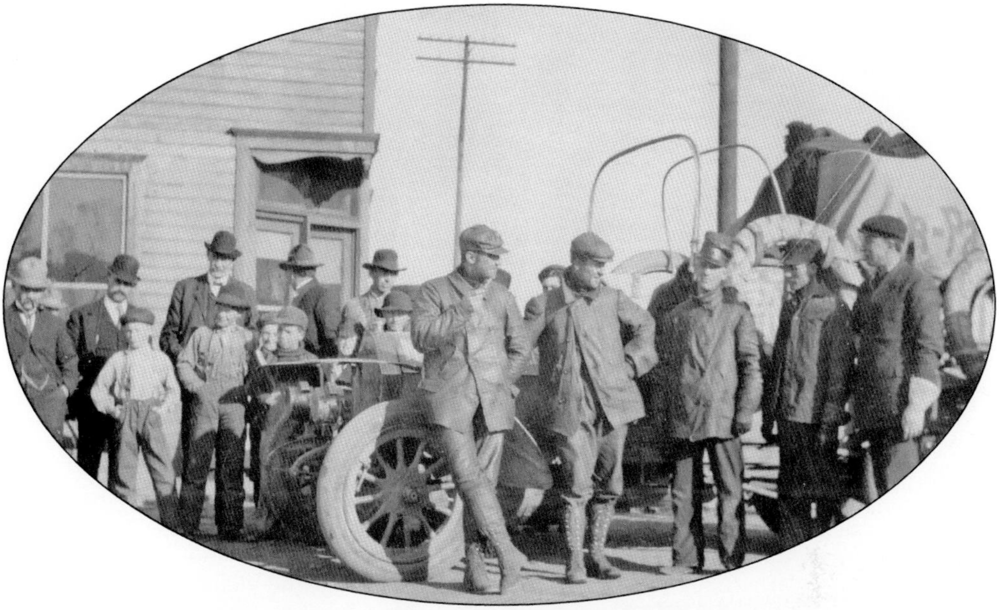

Between 1898 and 1908, new automobiles were being tested to see how durable they were and how far they would travel. In 1908, a four-door touring sedan (above) came through Kemmerer on its way from New York City to Seattle, Washington. The group stayed overnight at the Elk Hotel. Kemmerer held its first automobile show at the American Legion Hall in April 1929. The first automobile to be owned in Kemmerer was a bus; it was bought by a number of local men in 1906 who had organized a company to operate it. The Oldsmobile bus operated between Kemmerer and the coal mining camp of Cumberland. The second car (below) belonged to Dr. C. D. Stafford and the third to Patrick Quealy. Dr. Stafford is believed to be one of the men pictured below.

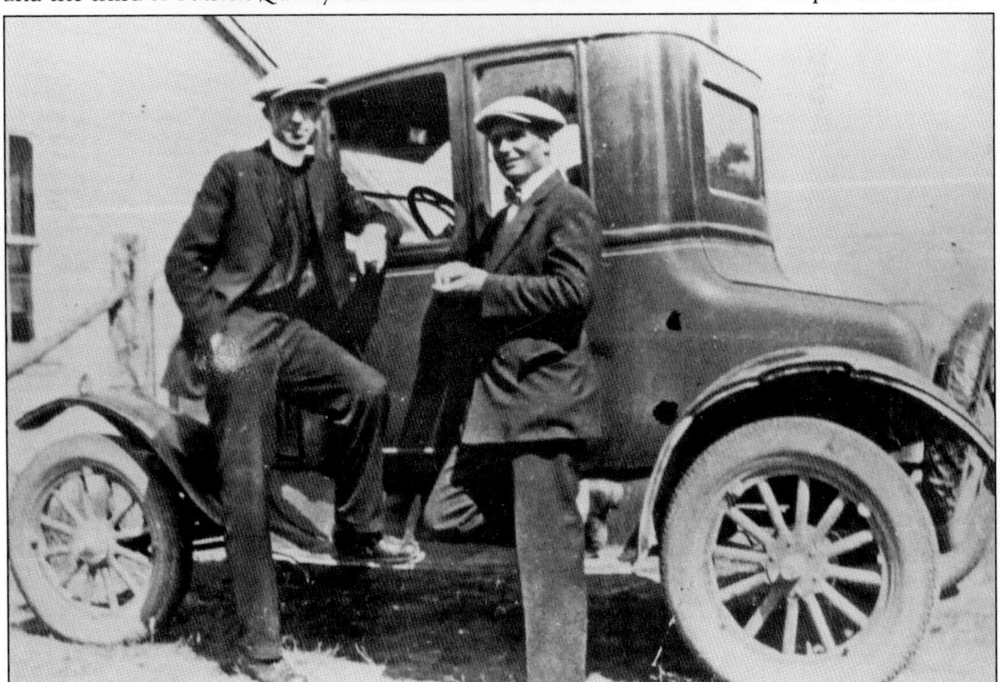

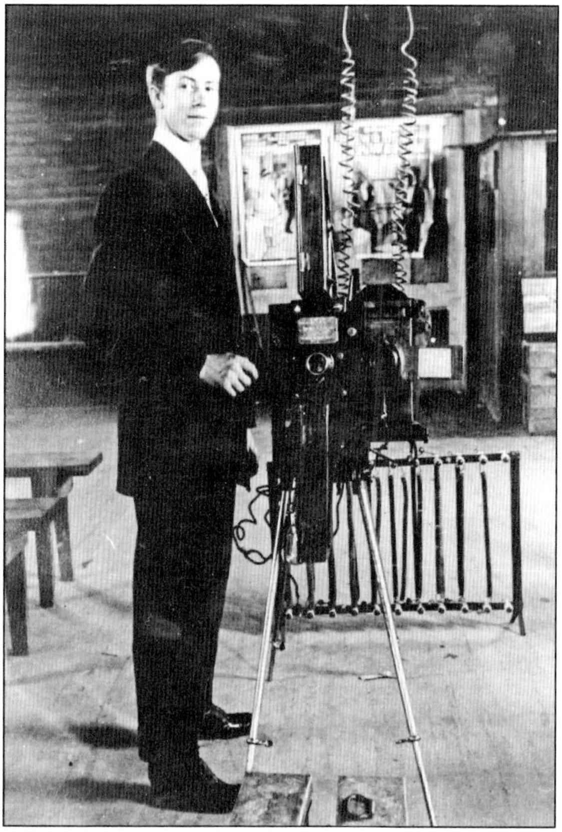

The Victory Theatre in Kemmerer was built in 1925 with funding from Quealy. The theater was owned by Willford Williams (left) and Frank Davis. The new theater was built on Pine Avenue, replacing a taxi stand. Williams' first job in the theater business was in 1909 when he ran movies for the coal mining camp of Cumberland. Throughout the years, the Victory was more than just a movie theater; it served the center of community life. It hosted talent shows, minstrel shows, weekly vaudeville acts, fiddlers' contests, dance recitals, and the state American Legion convention. The annual high school commencement exercises were also held here.

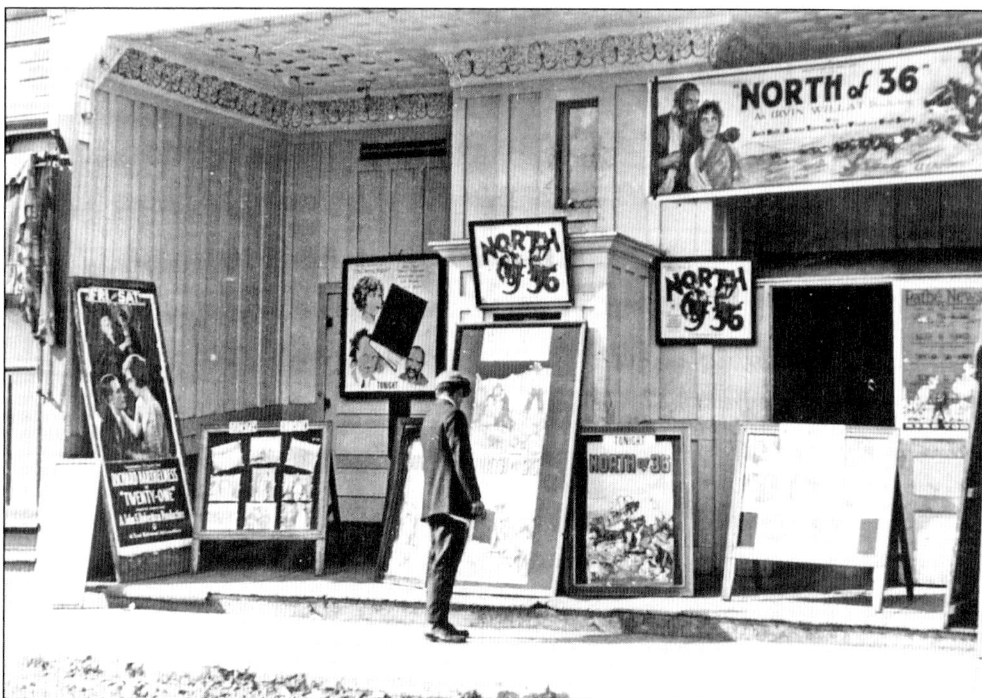

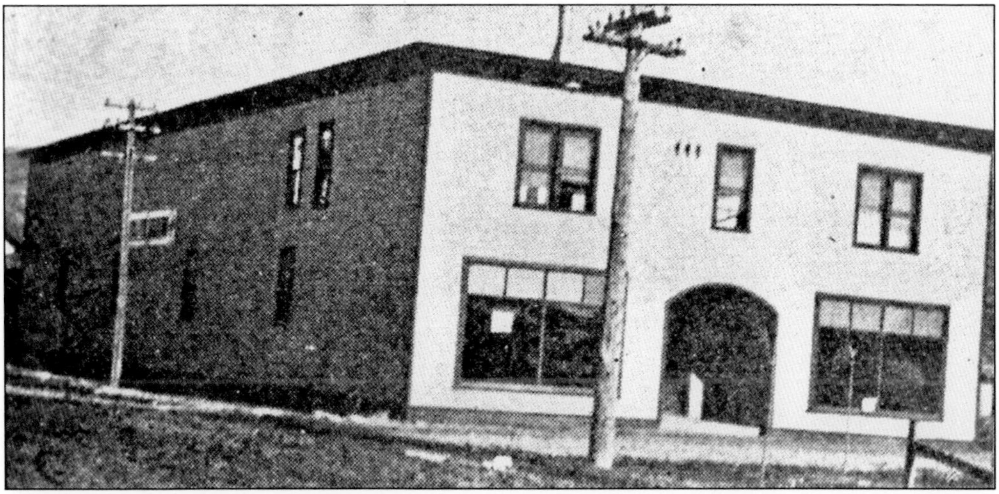

Cook's Pavilion was built in 1916 by Moses Cook; it was located on the corner of Sage Avenue and Topaz Street. The pavilion burned down in the 1950s, and now the bowling alley stands in its place. It was a roller skate rink and dance hall. In the fall of 1917, Vice Pres. Thomas Marshall addressed a large audience at the pavilion. The building was also rented out for large events like the Lincoln County Wool Growers' Ball and other large conventions. During the times the building was not rented, Cook brought in bands from all over the country to entertain the community. In 1917, during World War I, Company H was stationed at Cook's Pavilion.

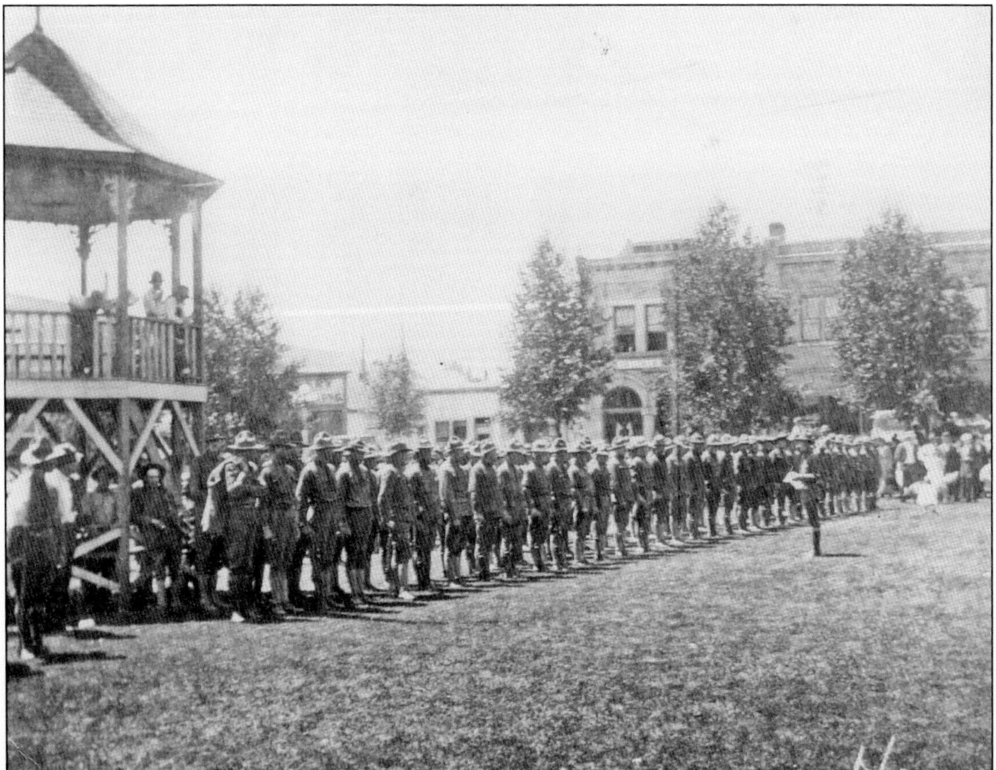

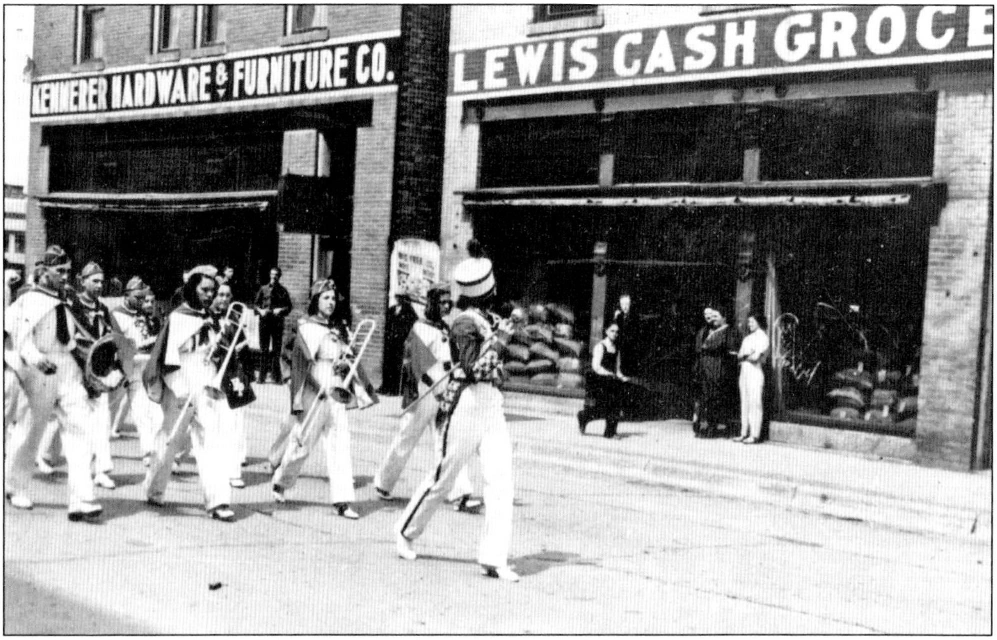

Band festivals held in Kemmerer began in the early 1940s and continued until the late 1950s. High school bands from Star Valley, Evanston, Green River, Lyman, Big Piney, and Pinedale came to Kemmerer for a two-day competition, usually held on Friday and Saturday. There were vocal and instrumental solos held in the high school auditorium all day. On Saturday, all the school marching bands lined up and paraded around the Triangle Park. Each was graded on its performance and marching ability. On Saturday night, the schools would combine their concert bands and chorus to present an all-school concert. The festivals were held in the spring, and the students and community always looked forward to this event.

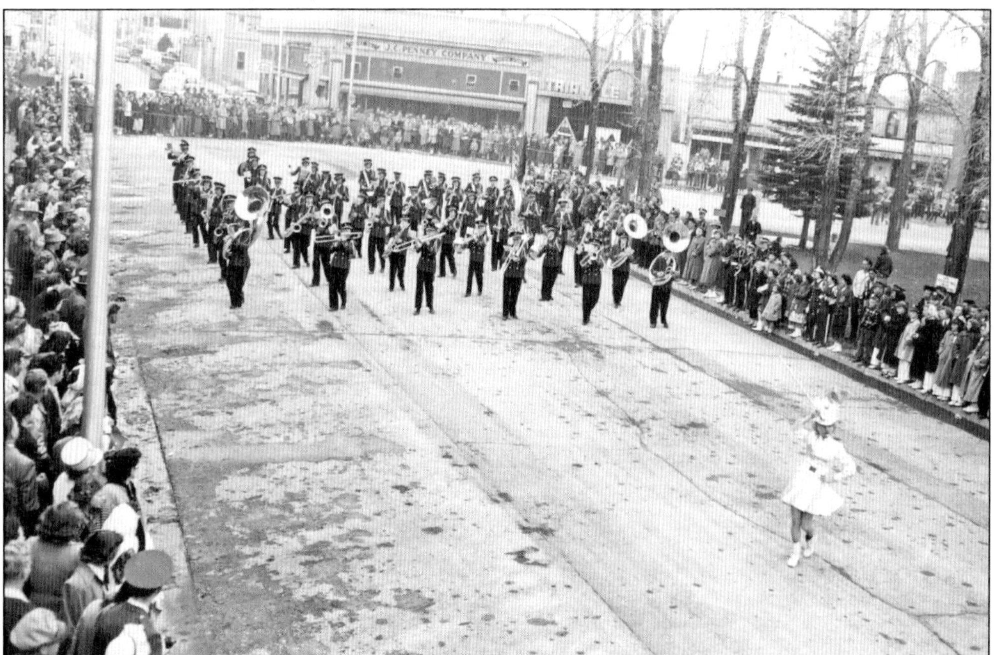

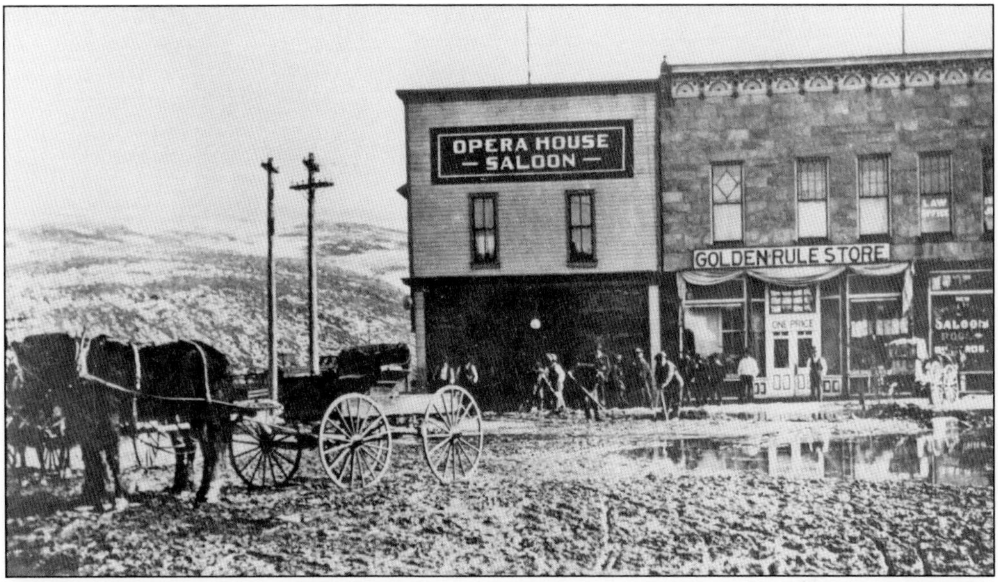

The Opera House Saloon, built in 1897, was the first dance hall in Kemmerer. It was first known as Keenan's Opera House, then Opera House, and later Casey's Opera House. The saloon was on the ground floor, and the dance hall and theater were on the second floor. There were minstrel shows, plays, and programs showcasing local talent. The school students would perform plays for their parents and the community in the hall, as pictured below. Fund-raising events for the school were also held in this building. Every Saturday night, a dance was held for the community with the music coming from local bands. In the spring of 1902, the Kickapoo Indian Medicine Show entertained, and in October 1902, a 20-round boxing match was held.

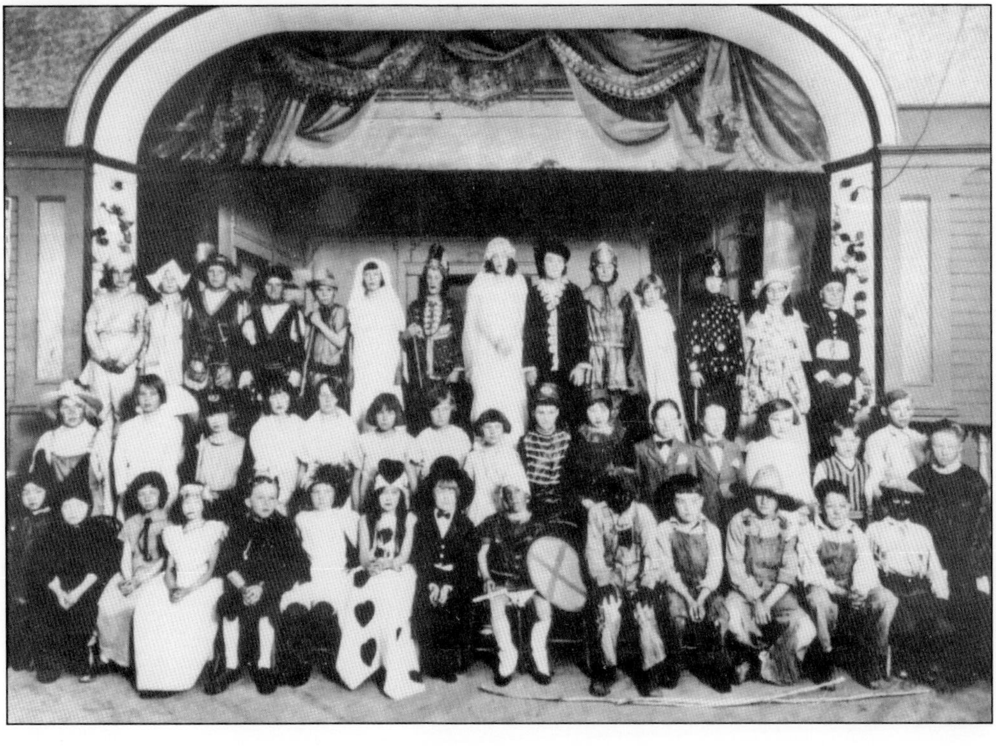

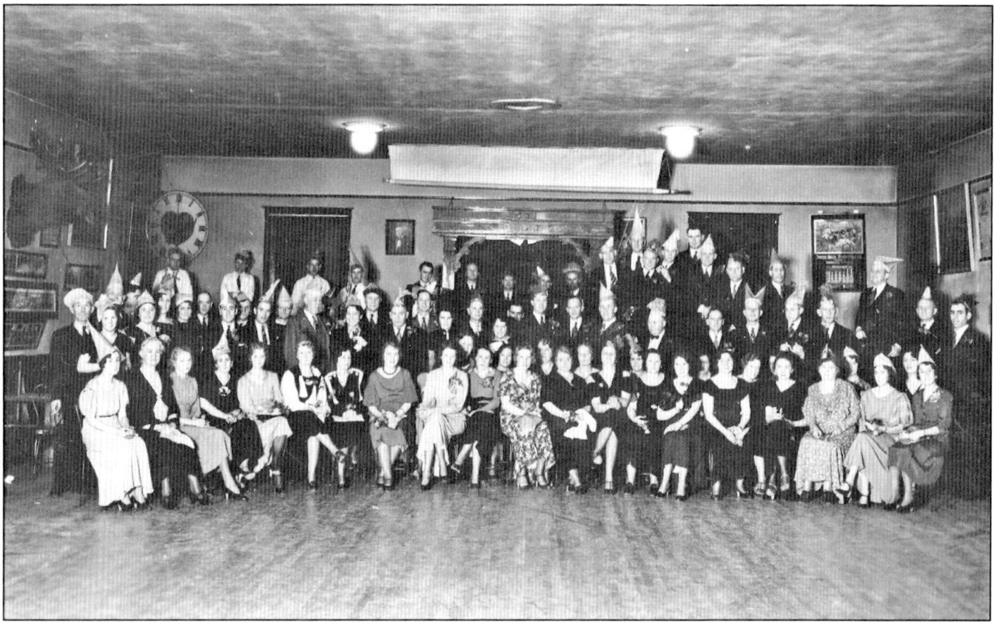

In the 1930s, dances and fancy dress balls were a favorite activity for the Kemmerer community. After many long hours of working, everyone was always ready for a party. New Year's Eve dances, Halloween dances, and many other celebrations were held in the IOOF Hall or the American Legion building. Music was provided by community bands. In the early 1900s, dances were held every weekend. Most of the surrounding coal mining camps had a place to hold dances. There was a dance hall at the county bridge on the Hams Fork, about 5 miles north of Kemmerer; this was a favorite place for the ranchers from the upper Hams Fork Valley. In 1929, the dance hall burned down and was never rebuilt.

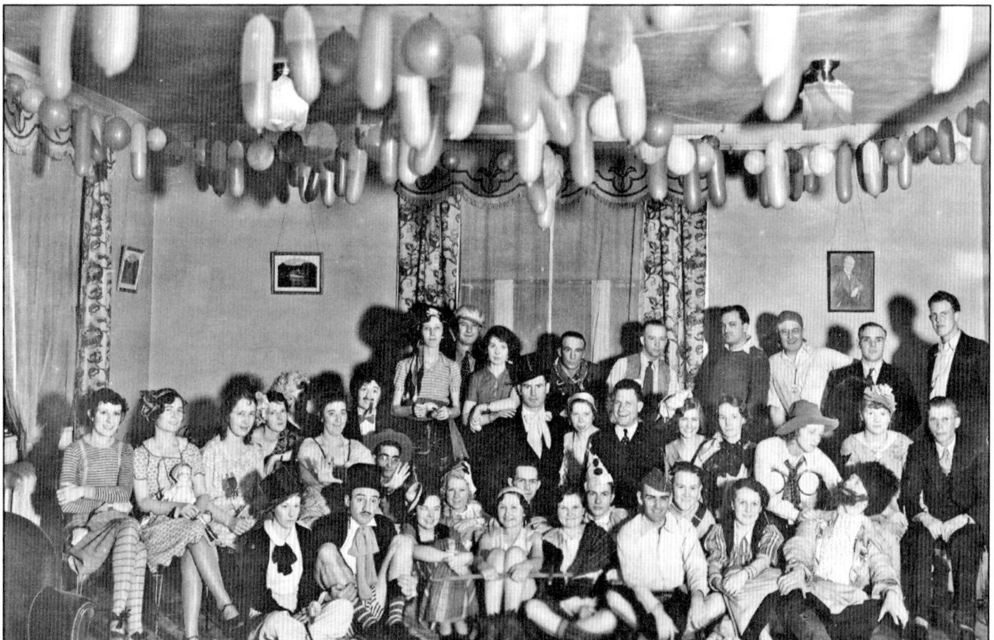

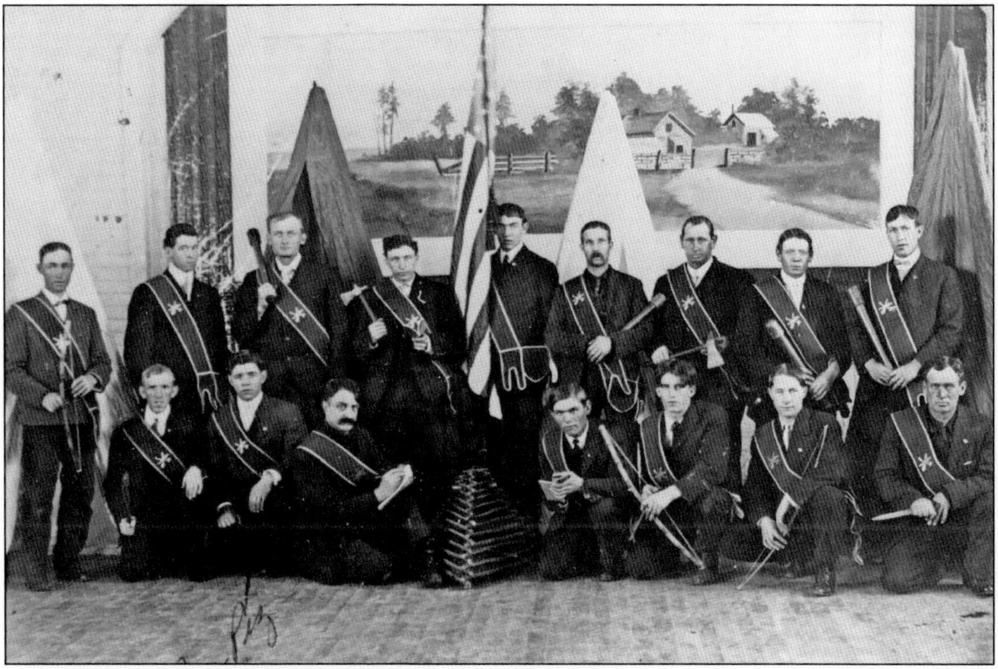

Kemmerer had four fraternal organizations by 1901. They were Kemmerer's Improved Order of Red Men, Odd Fellows, Kemmerer Fraternal Order of Eagles, and the Triangle Camp of the Woodsmen of the World lodge. These groups would meet in different saloons at least once a month. Later a Masonic Lodge was formed, first met with a lot of opposition. After a few years, most of the lodges had their own buildings to hold meetings or family events. Patrick Quealy was a member of the Elks, Knights of Columbus, and the Moose Lodge. Visiting brethren were always invited to attend the meetings. Pictured above is an unknown fraternal organization; below is the Moose Lodge.

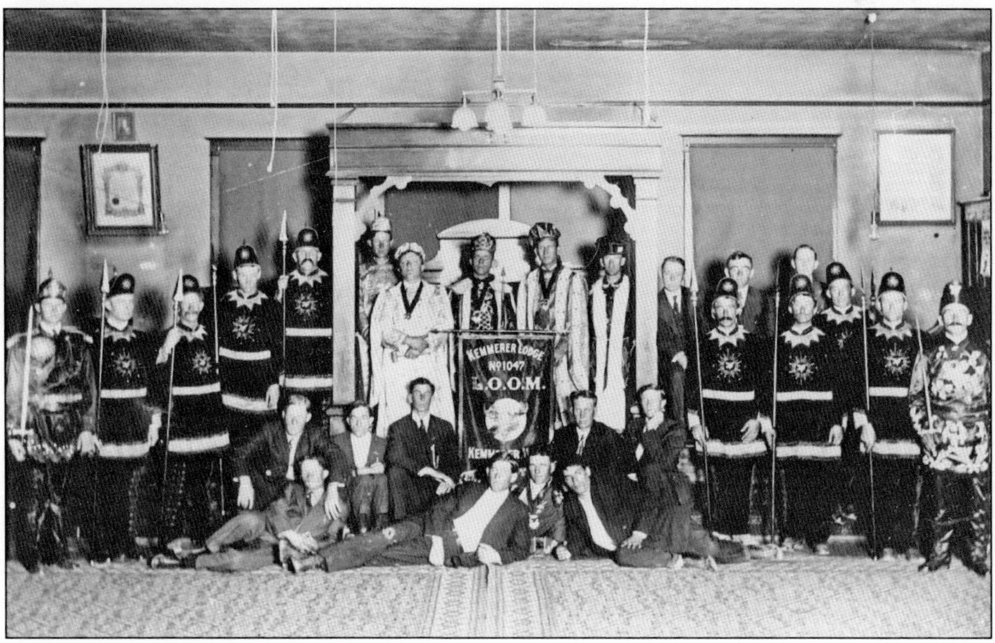

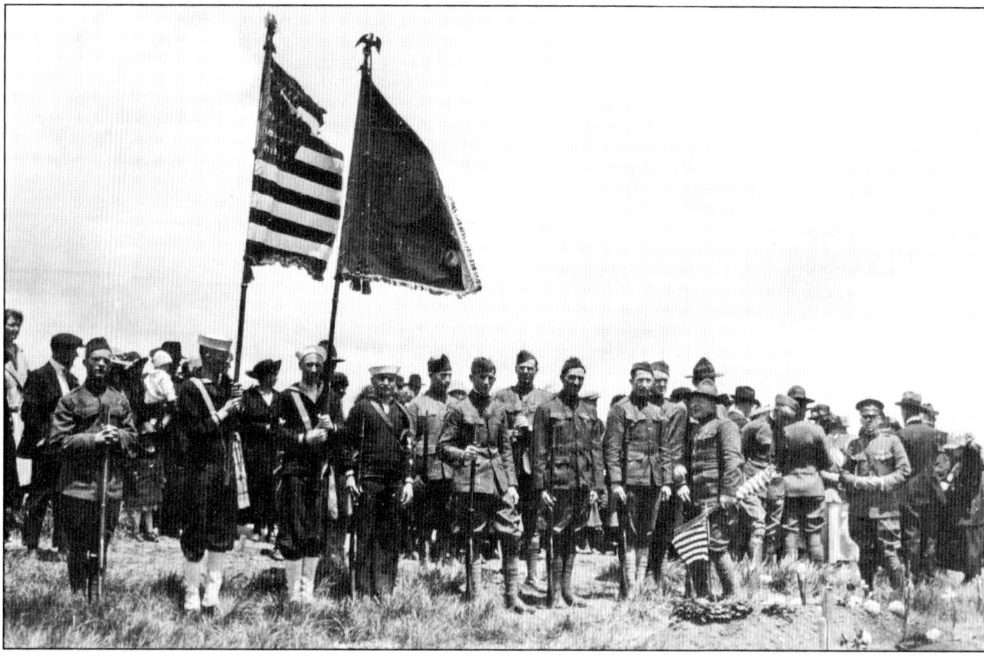

On Memorial Day, starting in 1922, the American Legion would march to the Kemmerer cemetery to honor the fallen World War I veterans. "Taps" would be played, a 21-gun salute fired to honor the fallen soldiers, and the American Legion members would place American flags on the graves. This practice continued into the 1950s.

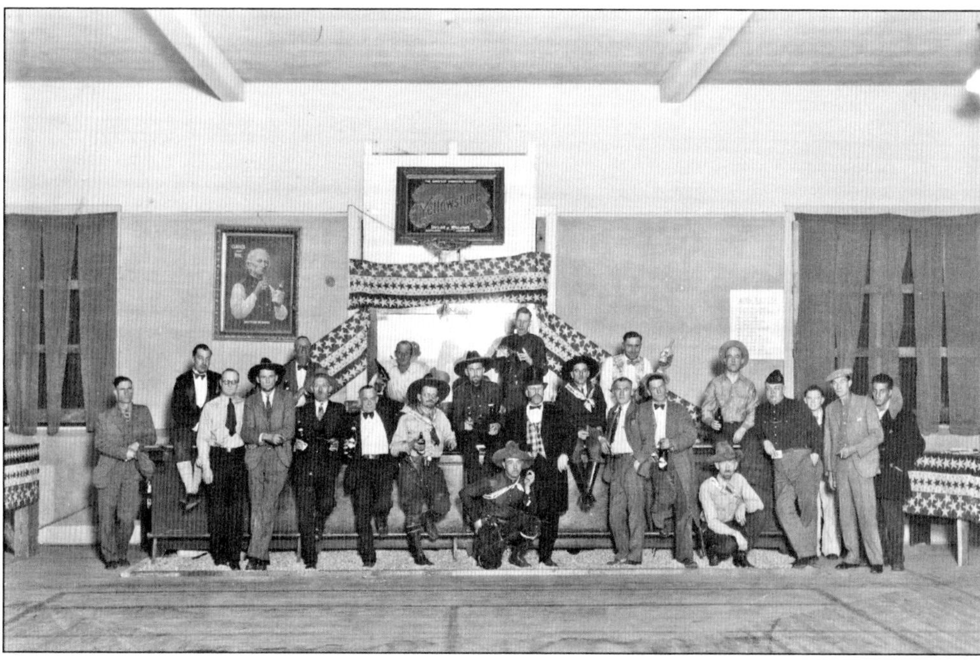

The photograph shows members of the American Legion, all veterans of World War I. The American Legion is a community service organization. It was chartered by Congress in March 1919. There was not a formal meeting place for the Kemmerer chapter until 1926, when the legion hall was constructed. The Kemmerer American Legion Auxiliary was formed in early 1922.

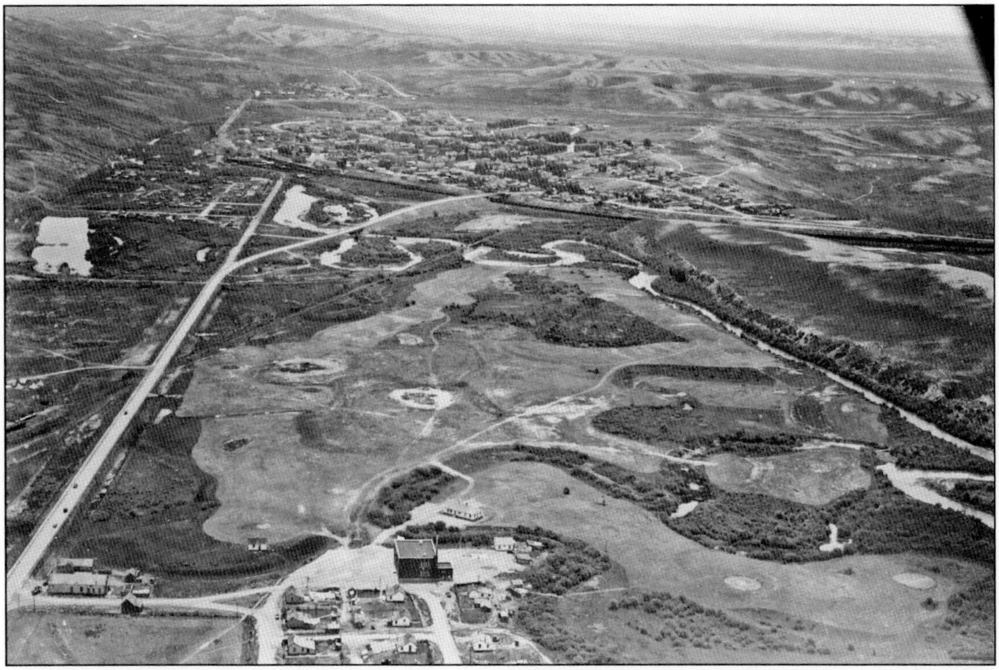

Kemmerer's golf club (above) was built on the homestead of Arthur Robinson; he had donated the land to the Kemmerer Coal Company for the town site of Kemmerer. A nine-hole golf course was built in late 1926, and the greens were ready for play on Memorial Day 1927. The golf course was named the Sagelawn Golf Club. The clubhouse was donated by John L. Kemmerer. This golf course was considered the best in the state. In 1941, grass had been planted, and keeping livestock off the course was a big problem. In 2006, the Kemmerer City Council approved the new name of Fossil Island Golf Club. In the early 1920s, the area children enjoyed ice skating on the Triangle Park. The fire department would flood the park and small fires would be built so the skaters could warm up.

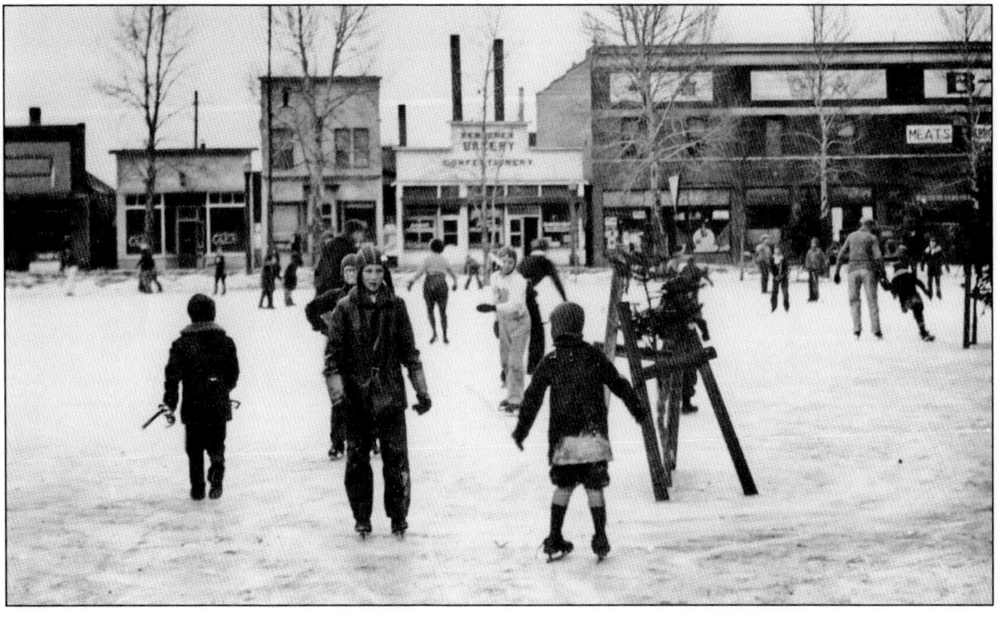

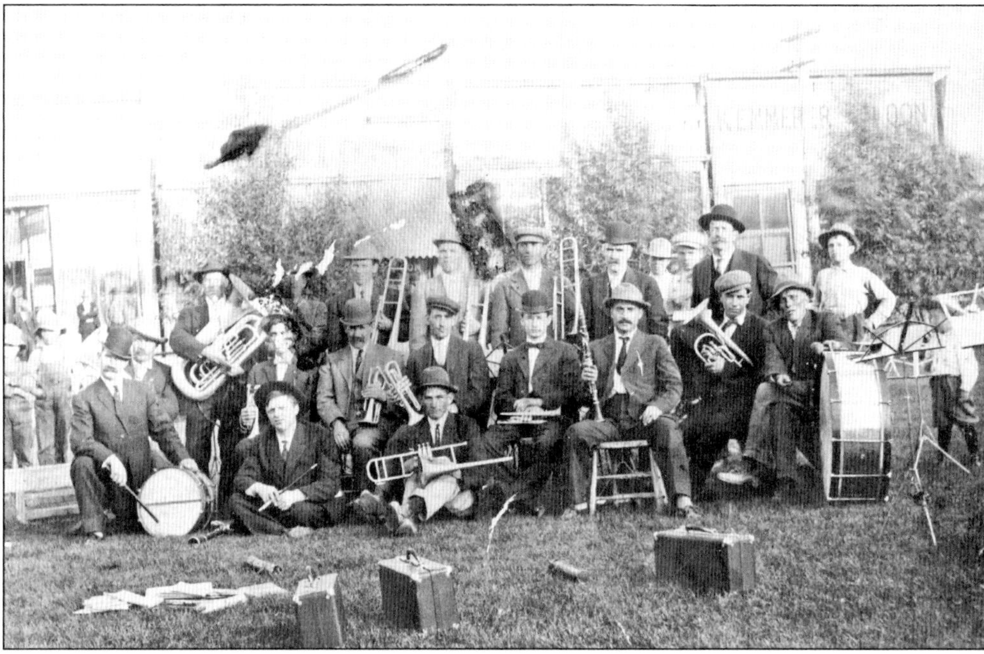

Kemmerer and the surrounding coal mining camps all had bands. Many of the self-taught band members were coal miners. Just like baseball, if a man could play an instrument, he would get a job in the mine. The bands would perform at all the coal camp dances and were expected to lead funeral processions to the cemetery. They would play on Memorial Day and Fourth of July celebrations. Hood's Community Band was Kemmerer's favorite; other bands were the North Kemmerer Band, Pryde's Bagpipes, the Cumberland Band, and Harmonica's 16-piece Band. The earliest known band in the Kemmerer area, called Professor Coe's Orchestra, dates back to 1899.

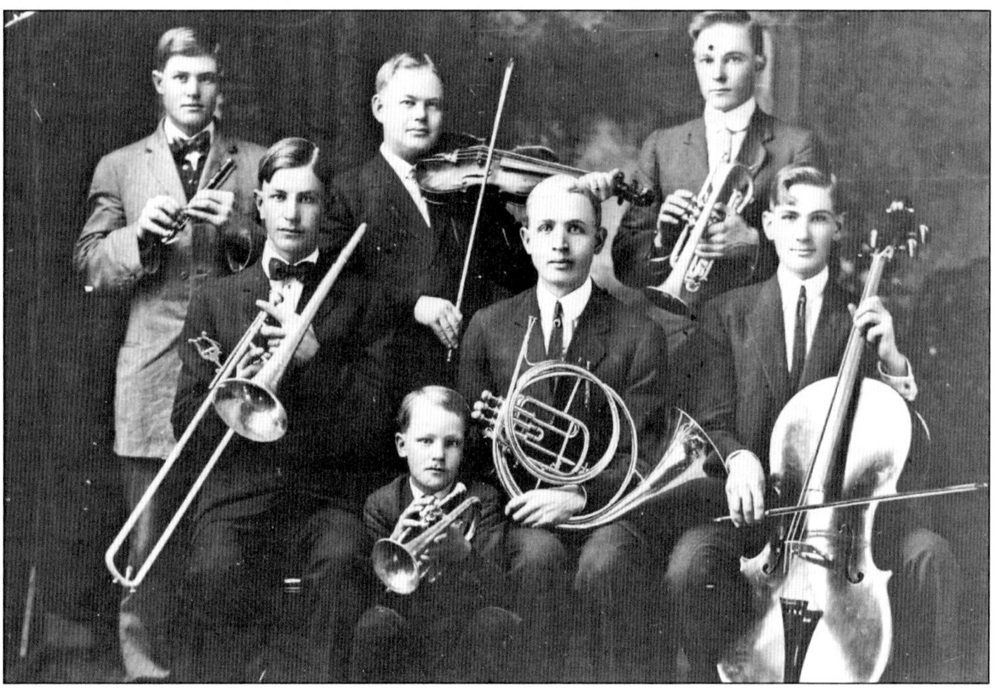

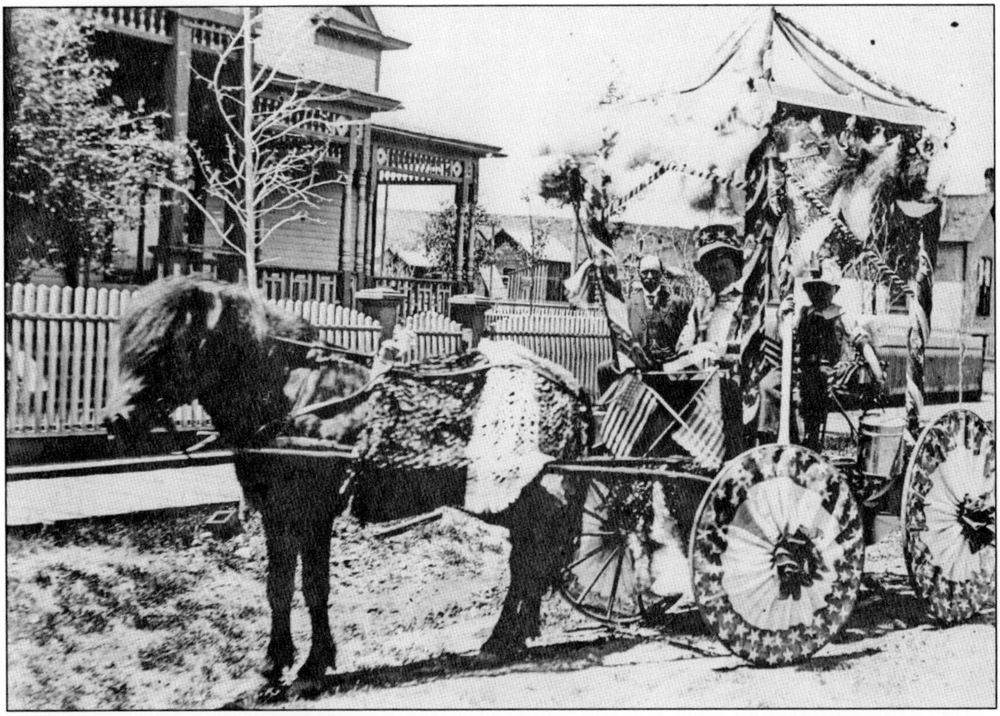

Parades have always been a favorite of small towns, and Kemmerer is no different. The parades started in Kemmerer as early as 1899. The first floats were horse-drawn carriages decorated with the theme of the parade. The most popular time to hold a parade was the Fourth of July. Almost everyone in town would be in the parade, especially children. This 1899 children's float (above) contains members of the Kemmerer Sunday school dressed in pretty costumes. A float from neighboring Diamondville (below) shows the Goddess of Liberty and her staff. Prizes were given to the best float. In 1940, the Old Timers Parade was held to raise money to help build the museum in the Triangle.

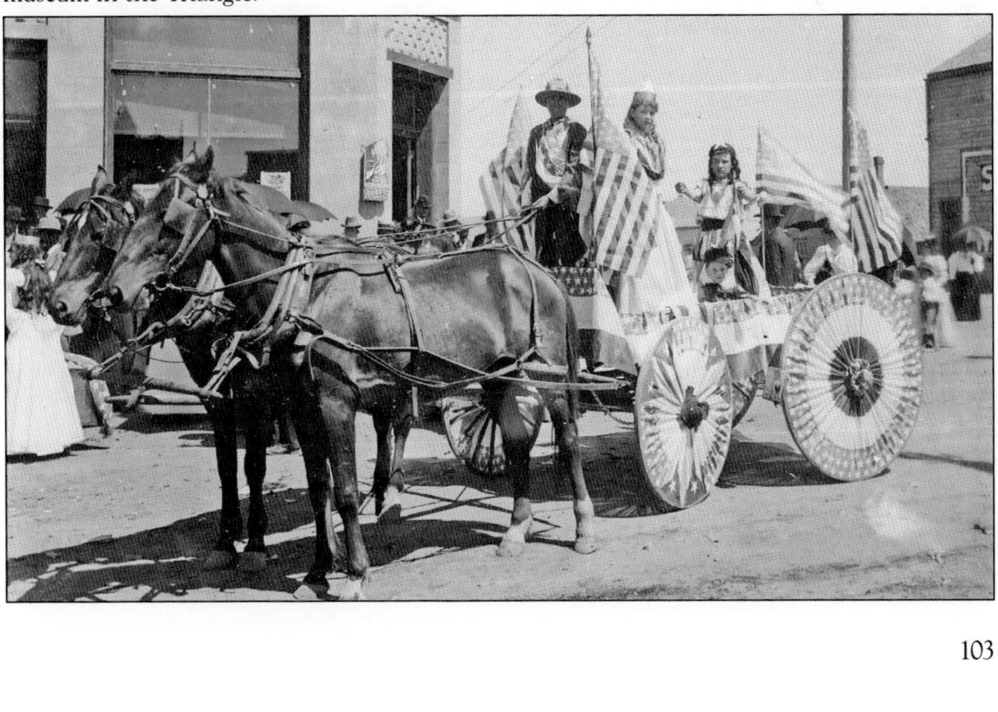

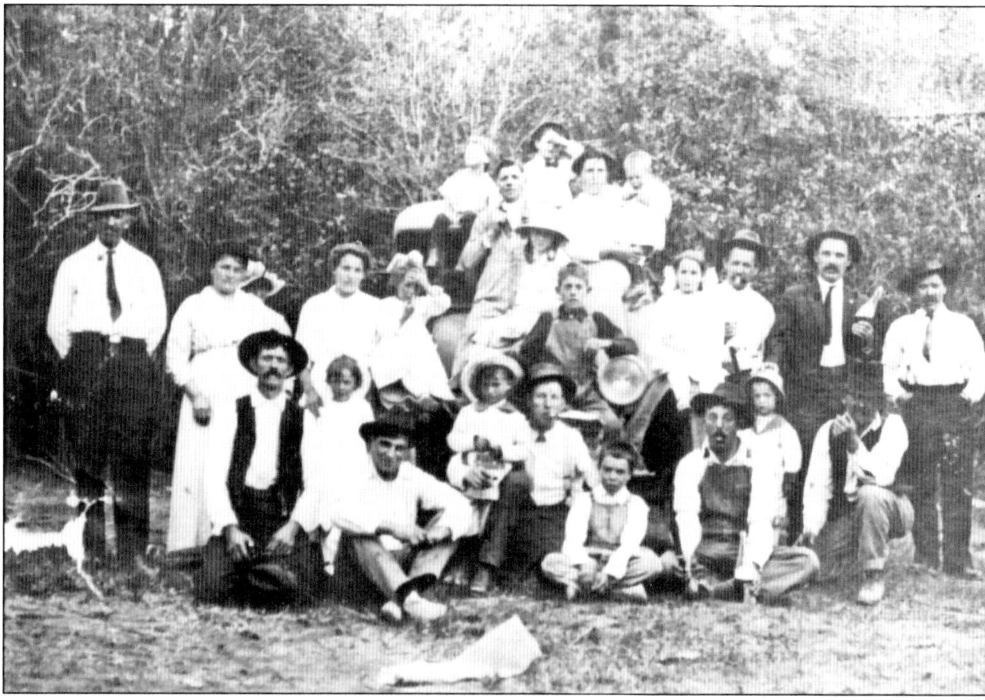

Many of the southern European coal miners who came to work in the area mines were from northern Italy and Austria-Hungary, which became Yugoslavia following World War I. The Italians and the Slovenes joined together for picnics or other social activities on occasion, but separate lodges or halls were maintained by each group. Of all the different ethnic groups in the area, the Finns, Slavs, Italians, and Scots were the most active for gathering for social reasons. In June 1902, the Finlanders of Wyoming celebrated the organization's annual Midsummer's Day in Kemmerer. Delegations from most of the towns along the Union Pacific corridor participated. Following a day spent along the river between Kemmerer and Frontier, several hundred people attended a grand ball held in the Frontier Community Hall. Pictured above is an unidentified group of European miners; below is a gathering of Austrian miners.

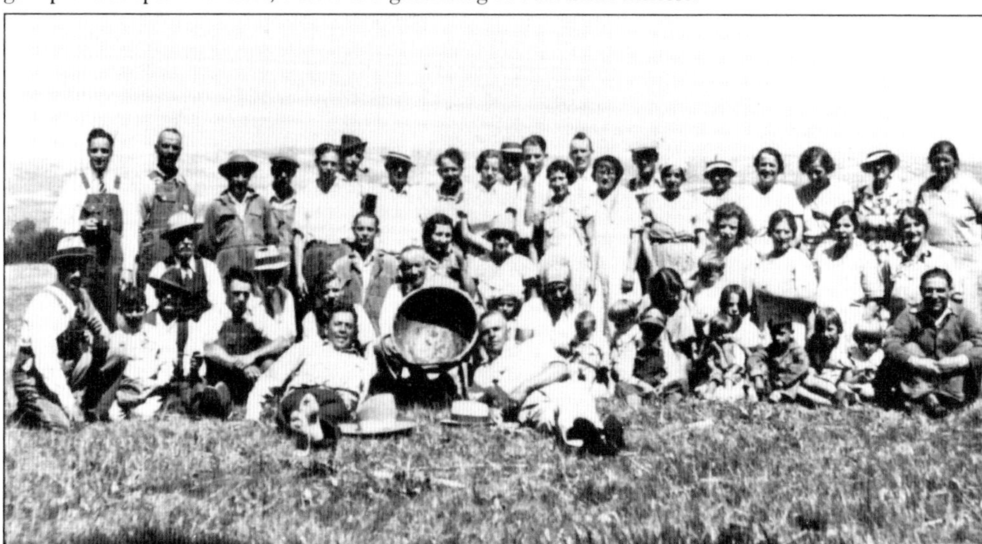

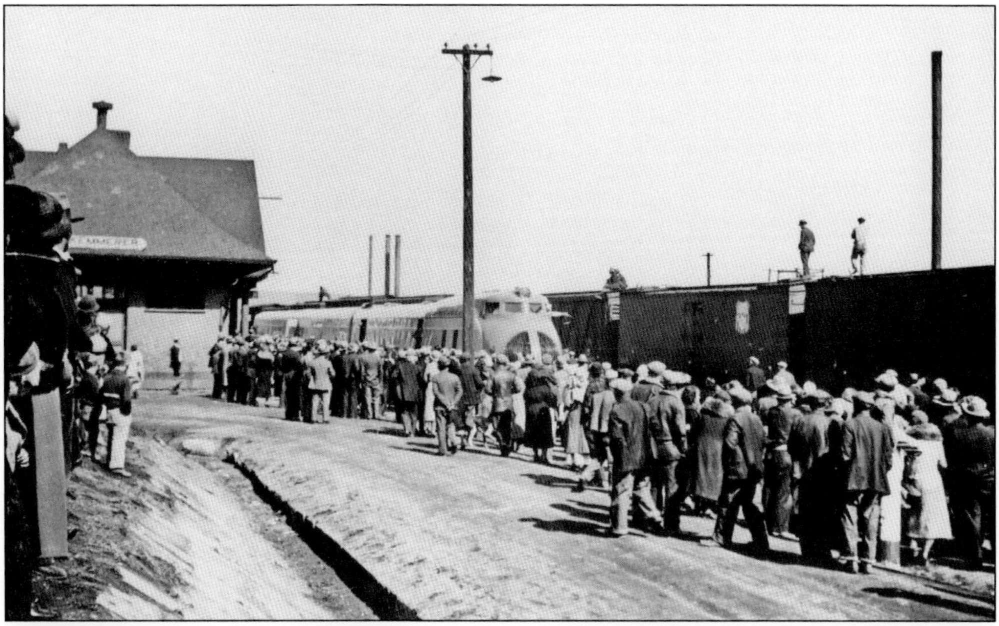

In 1934, the first steam liner that belonged to the Union Pacific Railroad came through Kemmerer. As pictured above, people gathered from all over the region to watch the train as it made a brief stop at the Kemmerer railroad yards to get water. The stop allowed the gathering crowd a chance to get a good look at the new engine. Below on May 28, 1989, the last steam engine built by the Union Pacific Railroad in the late 1930s came through Kemmerer. The engine was passing through on its last run from a centennial trip to the Northwest. These two engines most likely would have passed through Kemmerer during their first trips as well.

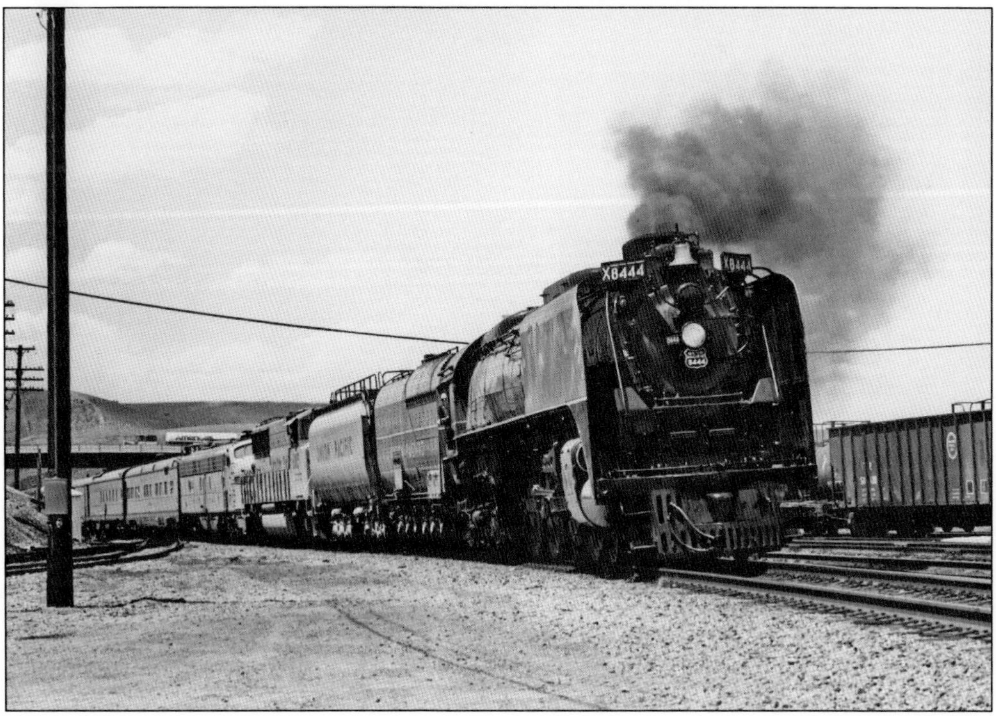

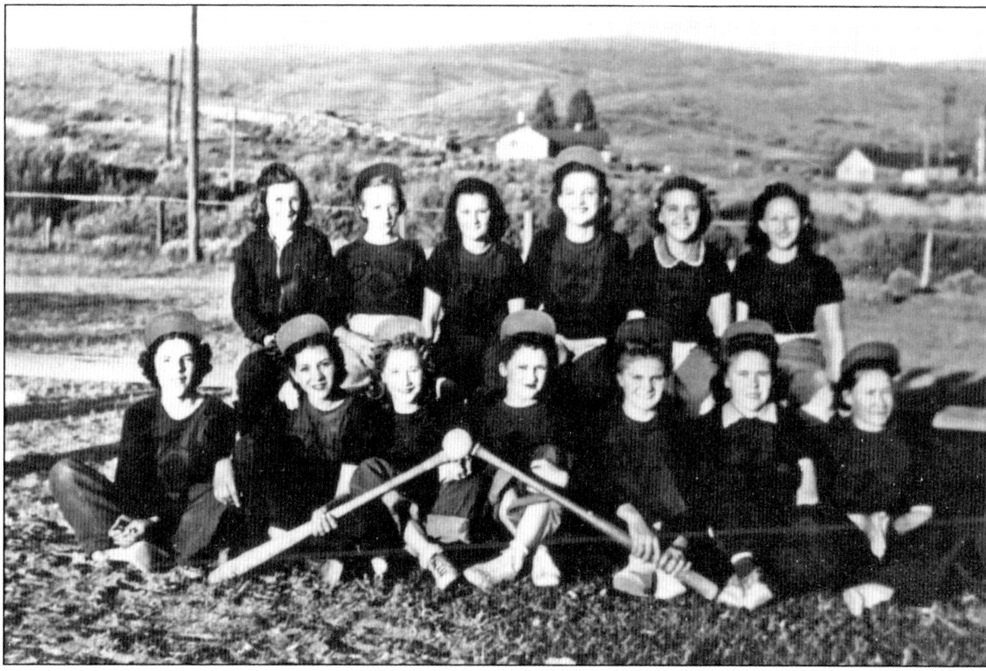

Baseball is not a sport just for boys. The Kemmerer area had its own girls' champion softball team in the 1940s. Women have been playing softball since the late 1800s, but the girls' teams gained popularity during World War II. Teams were formed starting in 1943 to keep the game of baseball alive.

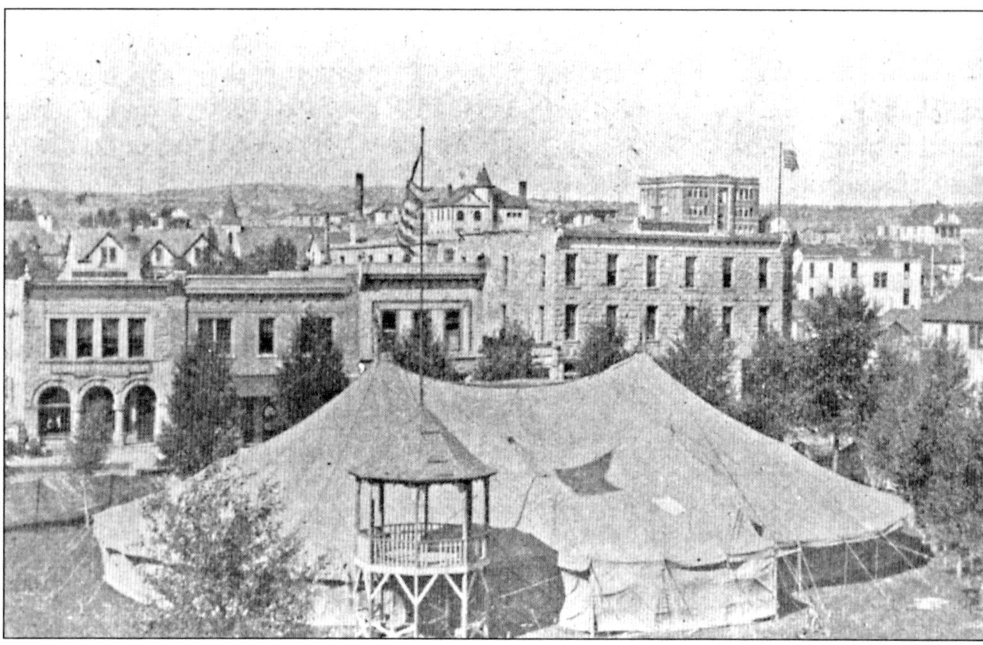

In 1917, many chautauqua traveling shows came to Kemmerer to perform. Chautauqua is a name given to traveling tent shows. The shows gave annual tours from 1907 until 1932. The large tents were always set up in the Triangle Park; the entertainment included ethnic musical groups, choirs, instrumental ensembles, and comedians.

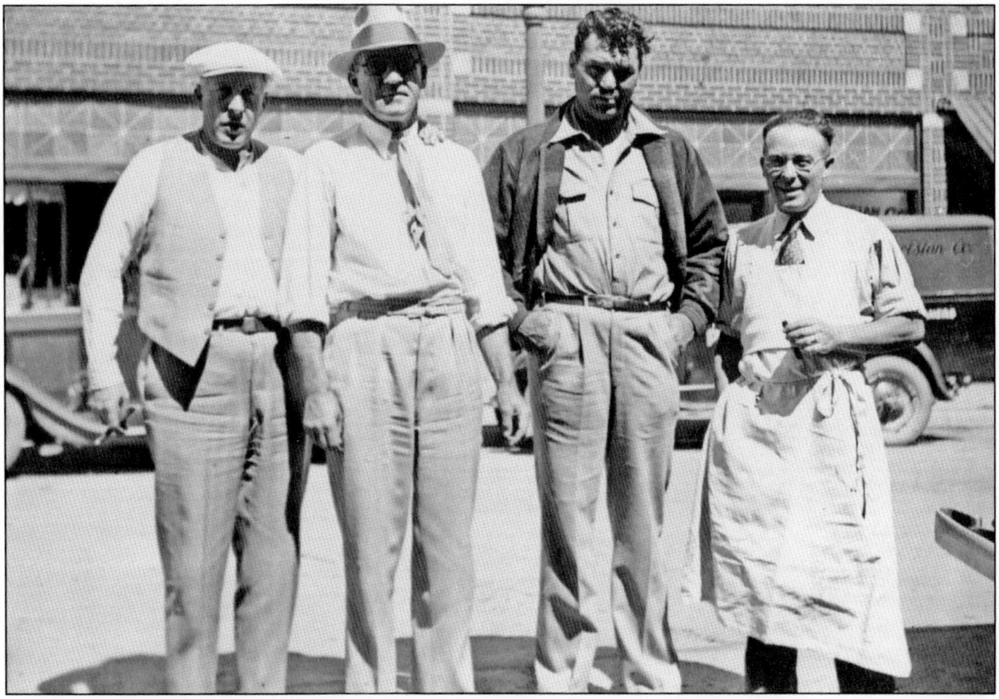

This photograph of heavyweight champion Jack Dempsey (second from right) was taken in 1928 in front of the Union Meat Market on Pine Avenue in Kemmerer. Dempsey came to the area to hunt sage chickens and had stopped at the store. In the photograph with him is Otto Angelo (far right), a butcher for the Union Meat Market. The other men are unidentified.

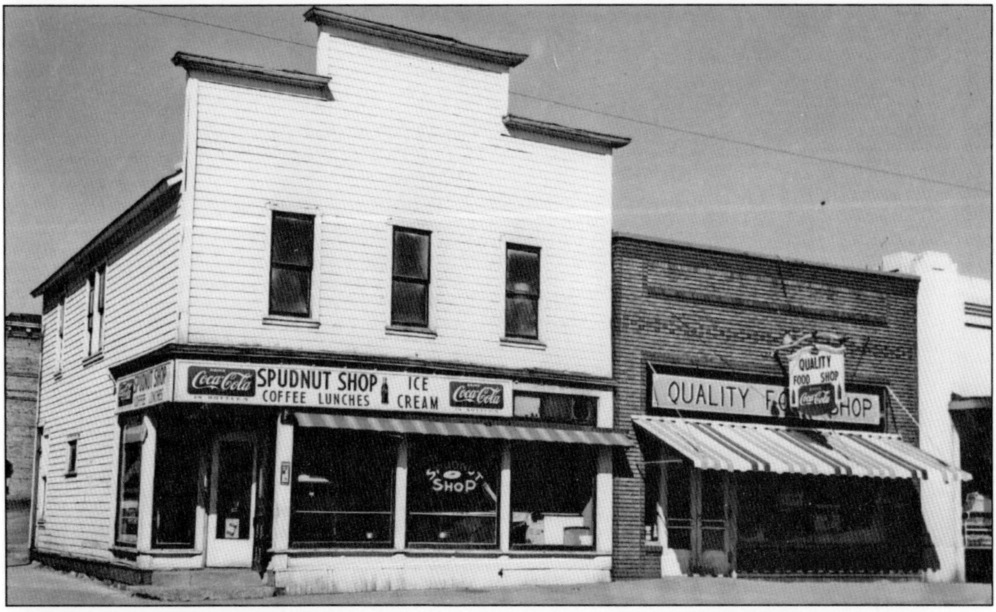

In 1950, the Spudnut Shop on Pine Avenue was the place to go for fresh spudnuts, hot coffee, and sandwiches. There was an art to making the spudnuts, and they were always good. The business was sold to the Picerno family in 1964. The business was moved to a new location on Pine Avenue. The family ran the spudnut shop and a restaurant, which closed in 1967.

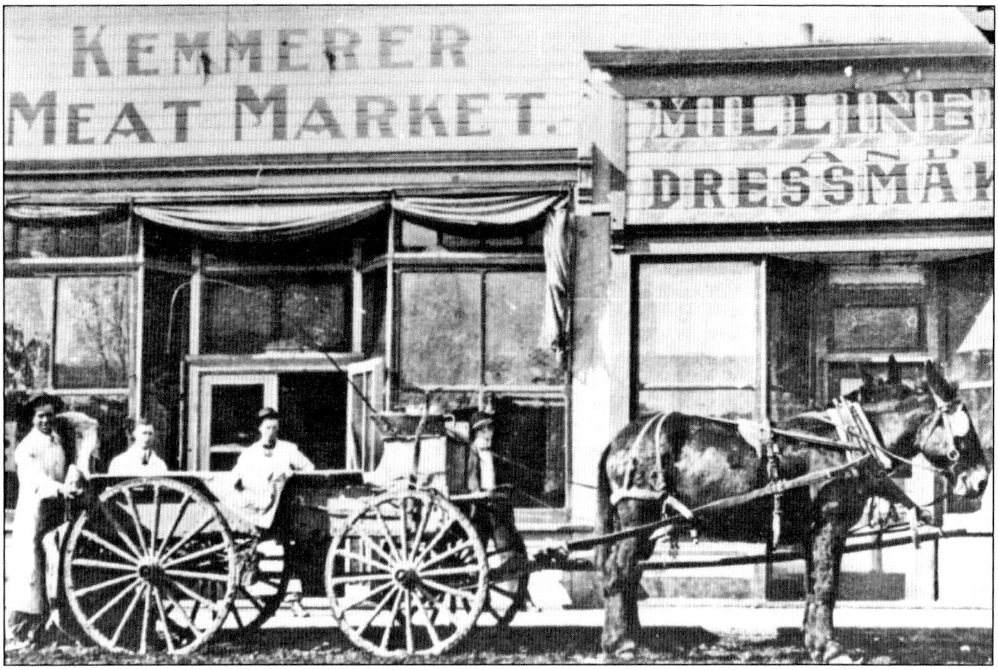

The Union Meat Market was opened in 1918 by the Peternal family. They ran the meat market until it was closed in 1974. The original stone building was constructed in 1898 by Luke Morris and Rufus Stanley. Mrs. Morris ran a millinery and dressmaking shop next door. The Sawaya family purchased the millinery building in 1909. Mr. Peternal learned the butcher trade in Austria while working for his uncle. When Mr. John Peternal first came to this country, he worked for Elkol coal mine, a Kemmerer Coal Company mine, before opening his store in Kemmerer. In the beginning, Peternal bought live animals and then slaughtered and butchered them at the Kemmerer Coal slaughter house in Frontier. He sold the meat in his store and made weekly deliveries to the coal camps.

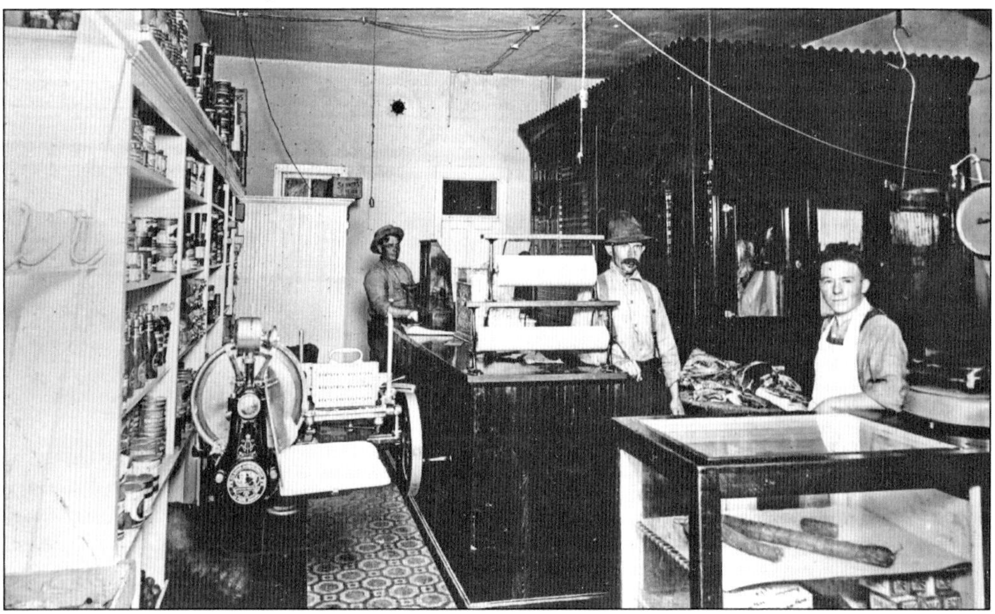

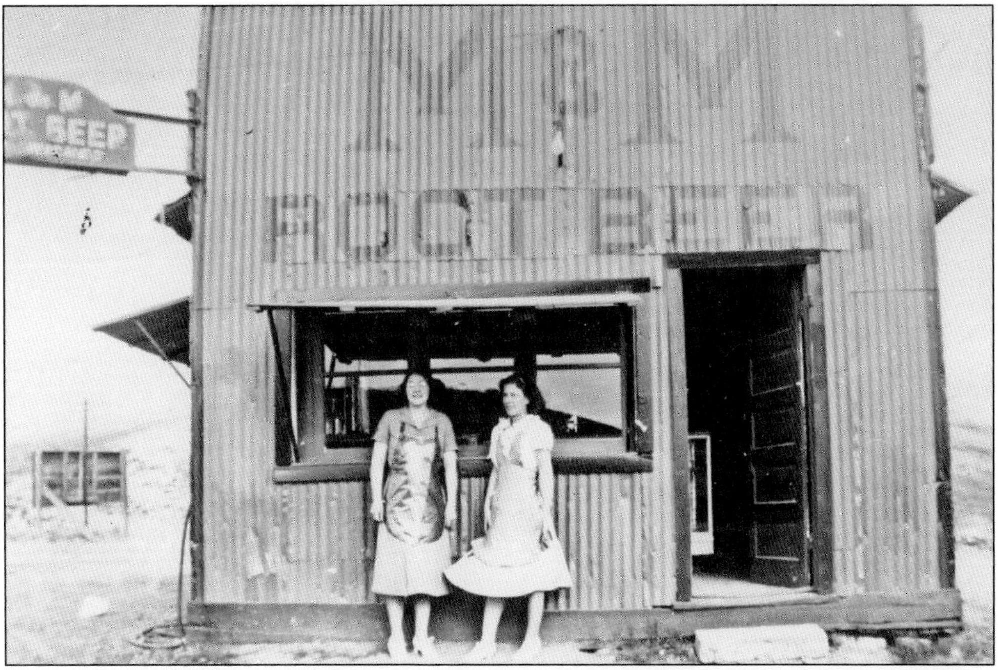

The M&M Confectionery was opened in the 1930s. The original location of the stand was on the south end of Pine Avenue. The girls from the Southern Hotel would walk up the hill to the stand to get a hamburger or shake. The confectionery was moved in the 1940s to a new location on Pine Avenue next to the First National Bank. It became a favorite teen gathering place. There was music and dancing, along with root beer floats, sundaes, soft drinks from a soda fountain, hamburgers, and French fries. Along with food and music, the confectionery also sold cigars, cigarettes, and tobacco.

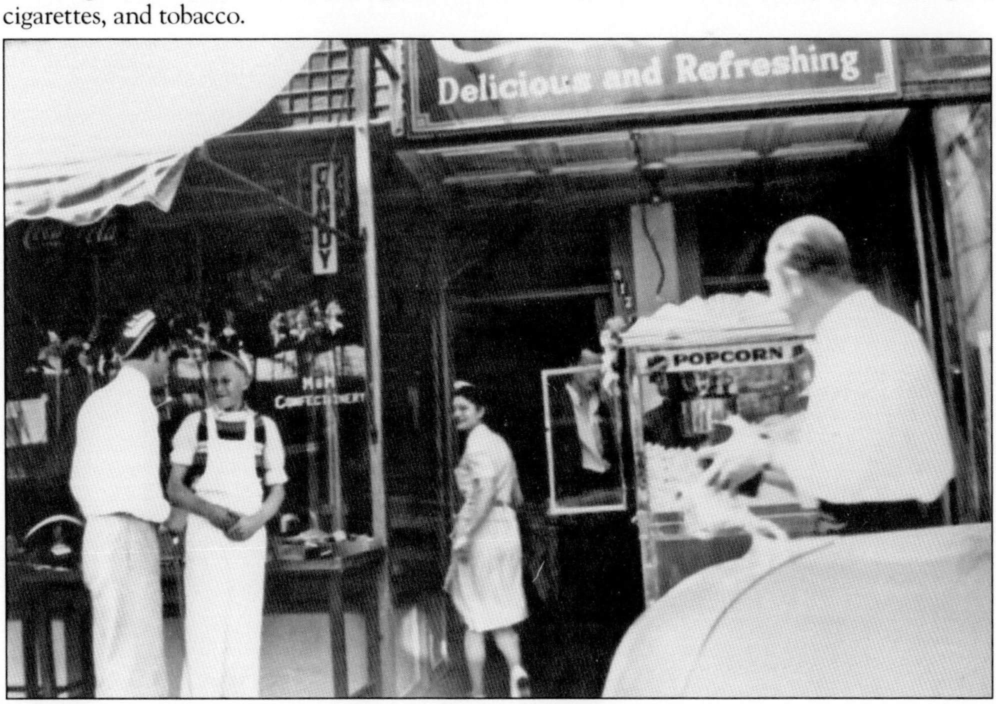

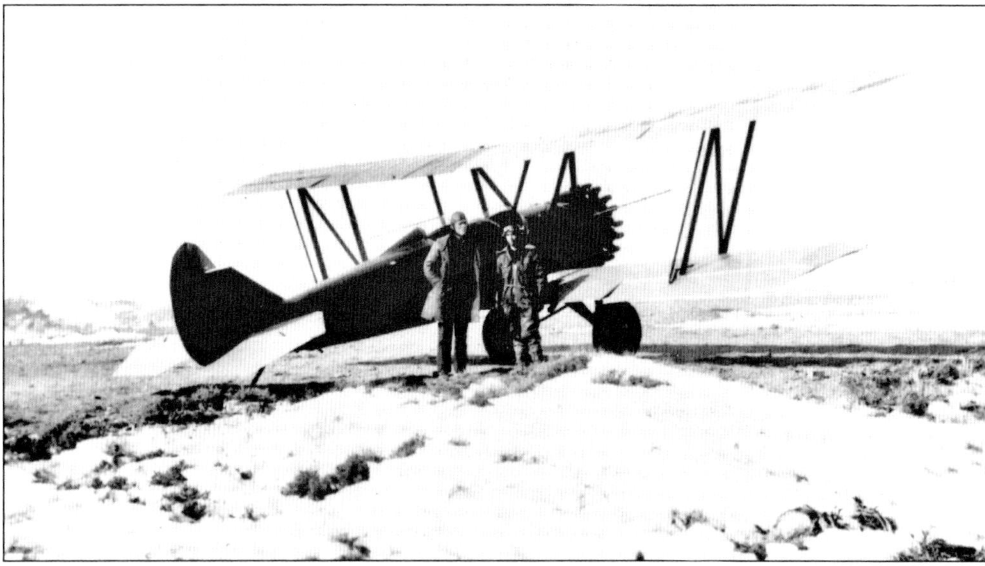

The first airstrips in the area were most likely flat hill tops, roads, or fields. The early pilots did not have the advantage of the modern-day airports. The Kemmerer Airport was built in the 1930s, at that time having only a single runway. The airport is located north of Kemmerer on a flat hill. At the time the airport was constructed, there were only two people in town who owned airplanes. In 1940, an airplane hanger was built to house the planes. The first outgoing airmail from the Kemmerer Municipal Airport was on May 10, 1947. In November 1965, the town council decided that the airport needed to be updated. The Kemmerer Municipal Airport has continued to improve and is frequented by several planes.

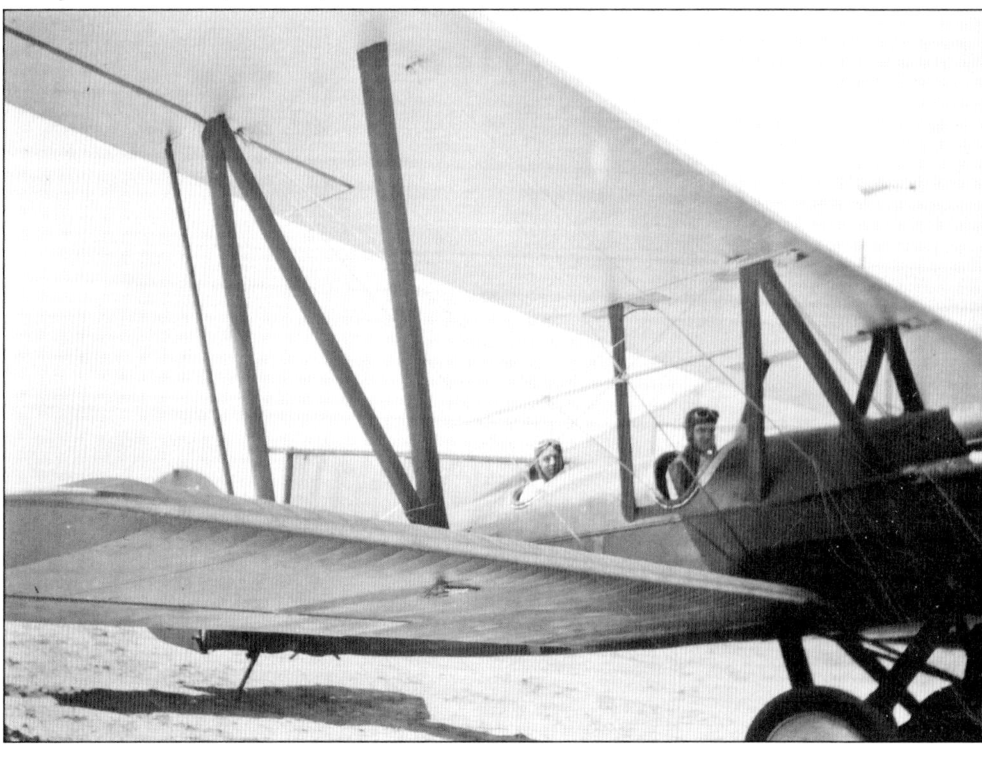

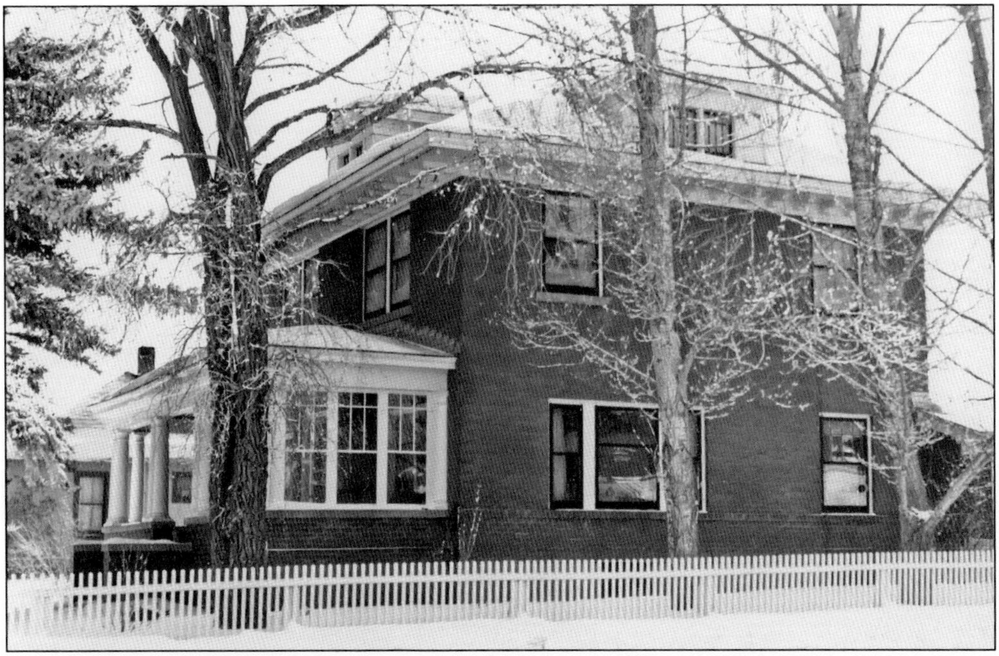

This home was built in 1900 by the Quealy family; it was later owned by Kemmerer attorney Robert Rose. Rose arrived by train in Kemmerer in 1914. He quickly became involved in the town's activities. He remained in Kemmerer for approximately eight years. The home was then purchased by John Reed, president of the First National Bank.

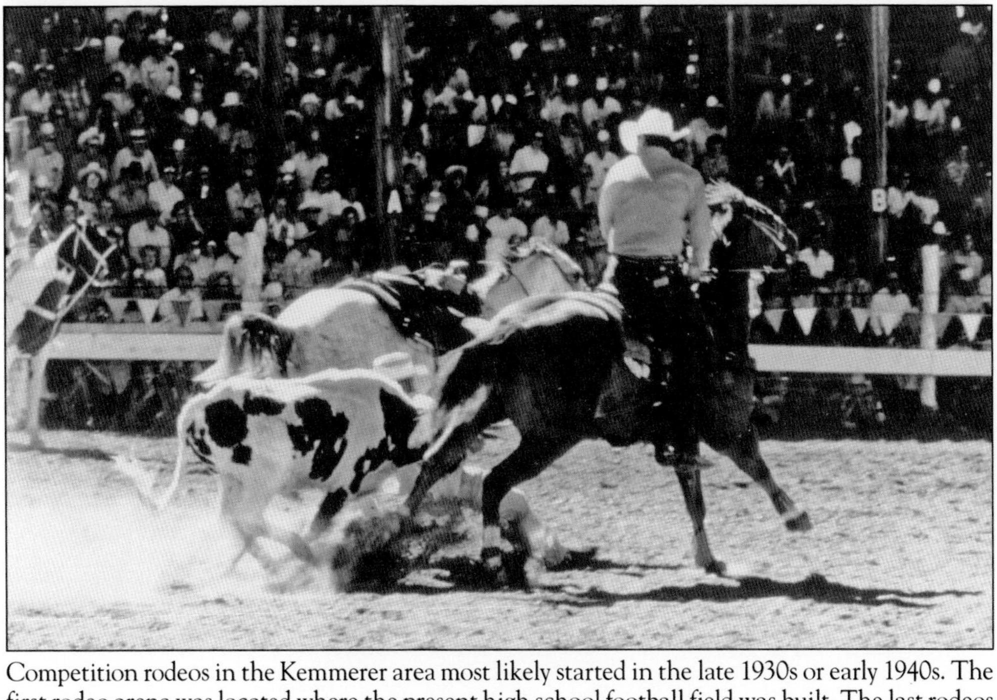

Competition rodeos in the Kemmerer area most likely started in the late 1930s or early 1940s. The first rodeo arena was located where the present high school football field was built. The last rodeos held at this location took place in the early 1960s. In 1971, Kemmerer again had a rodeo. Called the Big "K" Rodeo, it lasted only a few years. The arena was located north of Kemmerer.

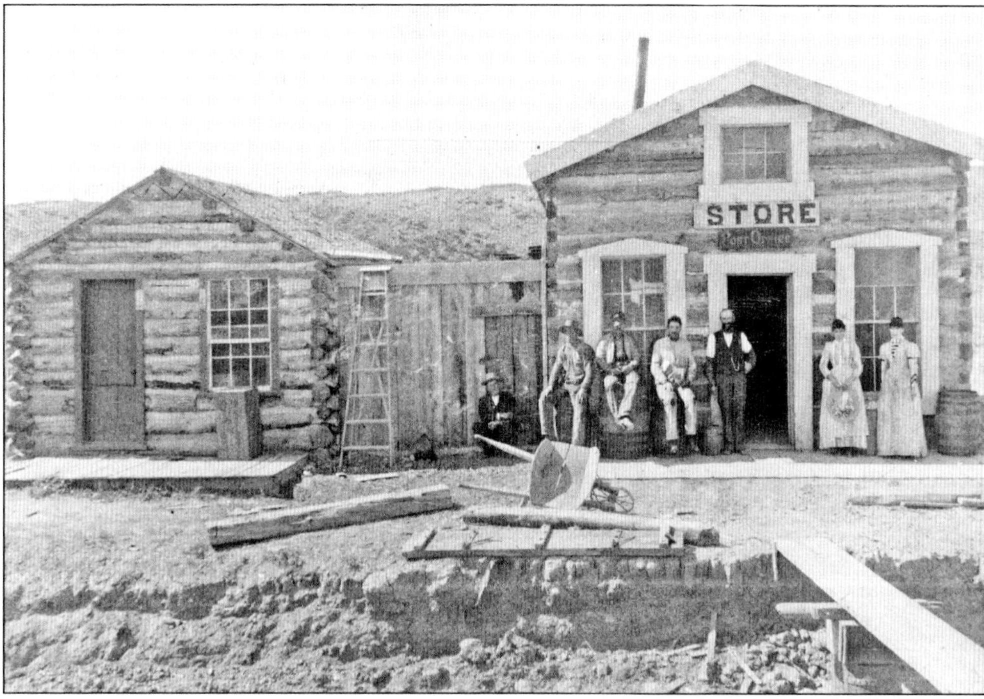

Located to the west of Kemmerer was the town of Fossil. It was started in 1881 when the Oregon Short Line Railroad was built through the area. The railroad built a station for servicing the trains. An engine house, coal chutes, depot, section house, and maintenance house were all located in Fossil. The coal for the trains came from the Twin Creek mine just a few miles to the east. Slowly the town of Fossil began to grow. By 1891, Fossil had a saloon, store, and post office run by Peter Gordon, who also built the Stock Exchange Bar in Kemmerer. Fred and Jane Cook also opened a restaurant in 1891. Jane Cook cooked sage chicken and fish for the men working on the railroad.

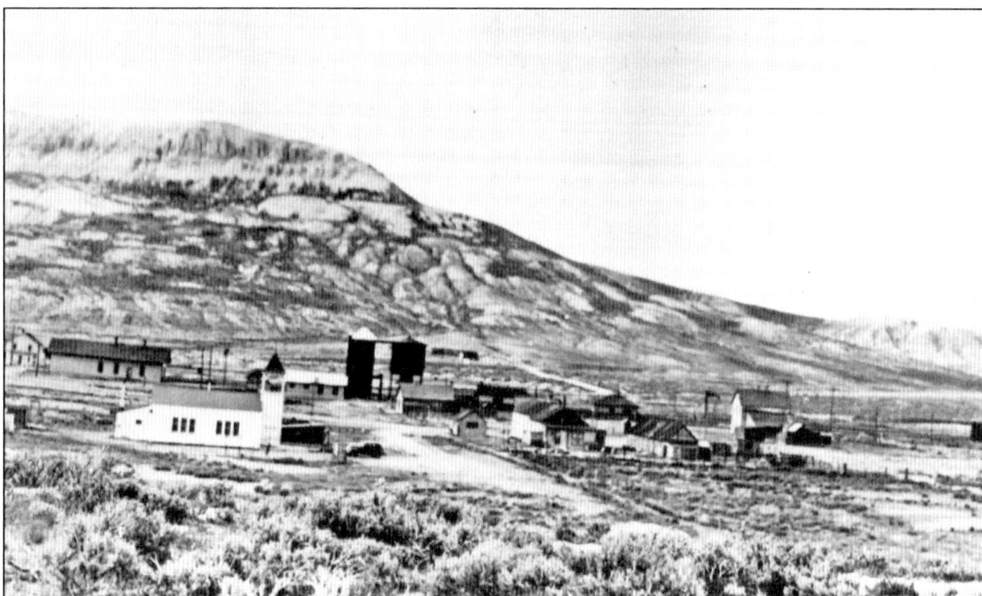

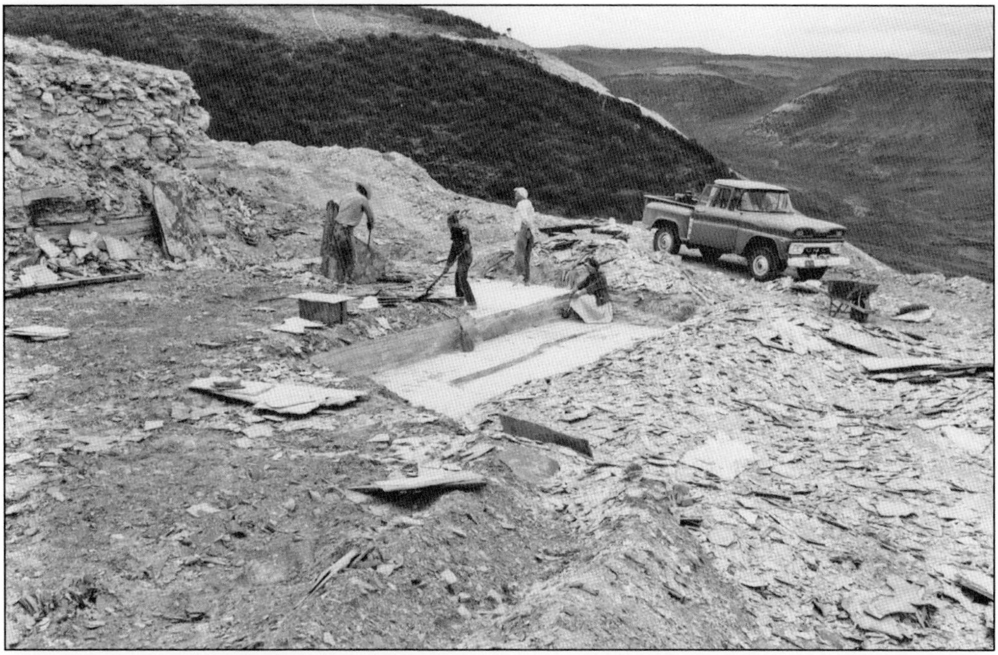

The coming of the railroad to Fossil brought hunters of all kinds, including fossil hunters. The earliest documented fossil hunters in the area were in 1897. The most colorful fossil hunter was Robert Lee Craig. He helped put Fossil on the paleontology map. Craig sent the fossils he found all over the world. Today fossil hunters can go to one of the fossil galleries to view or purchase the fossils or go to a quarry to dig for their own. The region is known as the Green River Formation, and this is what made Kemmerer the "Fossil Fish Capital of the World."

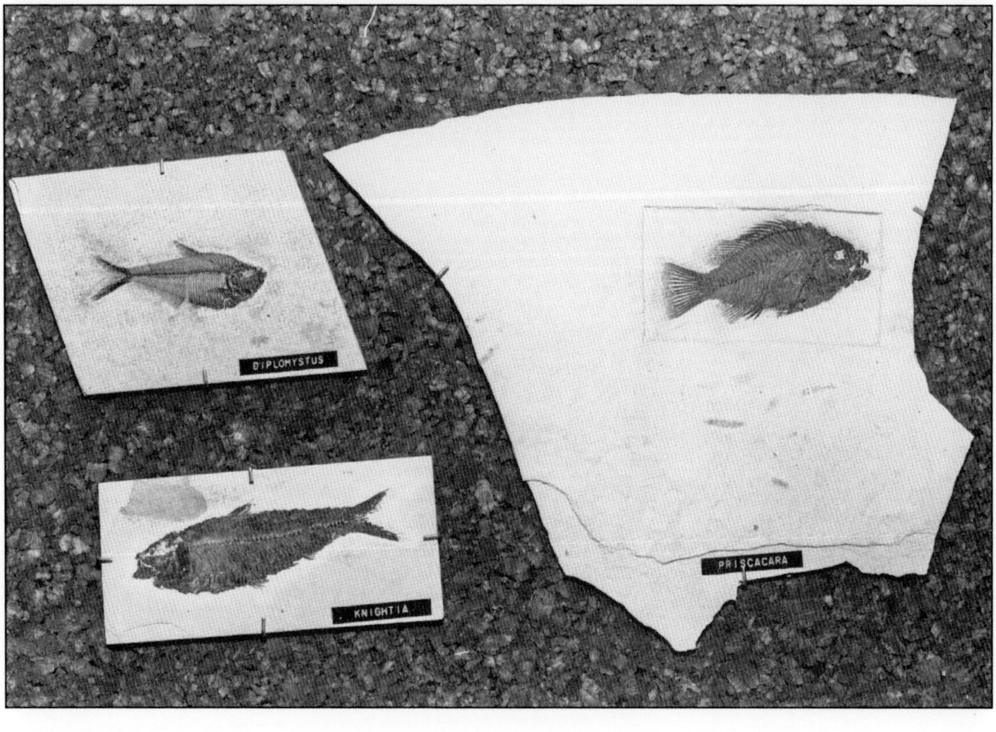

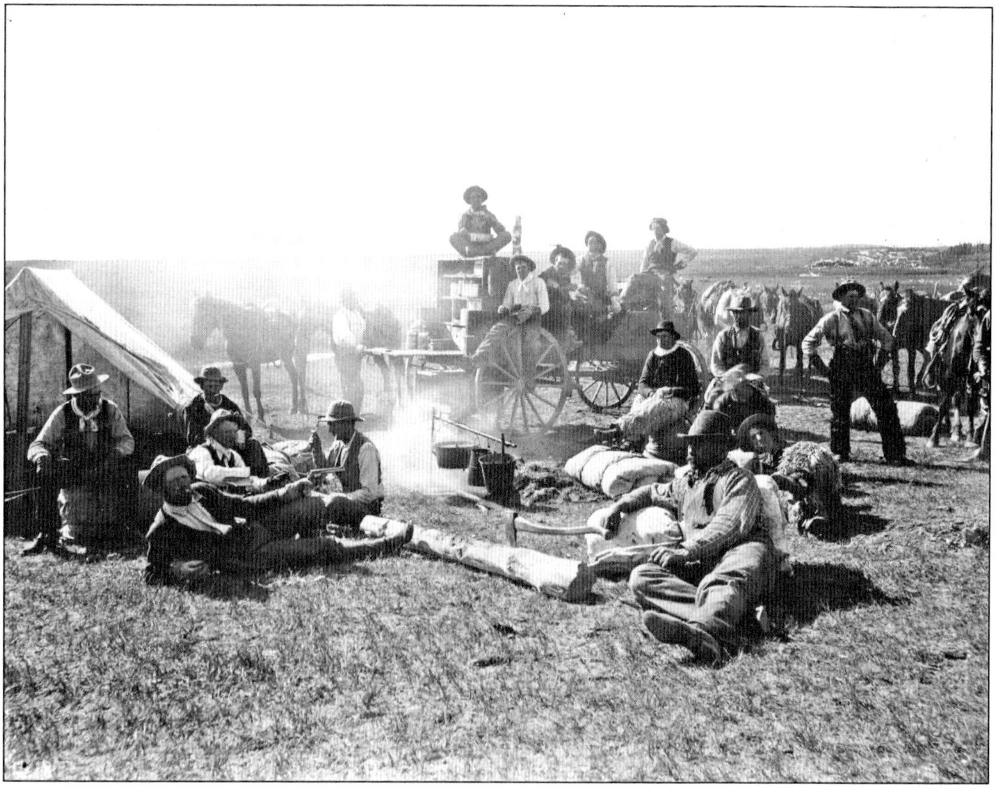

During the warm months, most of the coal mines were idle, working only a few days a month and with only small crews. The demand for coal in the warmer months was very low. The miners had to have some way to make a living for their families. Some were bootleggers, and others chose to go to work for local ranches. The miners would turn into cowboys and help brand the cattle in the spring, turn the cattle out to summer pasture and hay in the summer months, and when fall roundup would start, the men would help gather the cattle from the upper Hams Fork Valley.

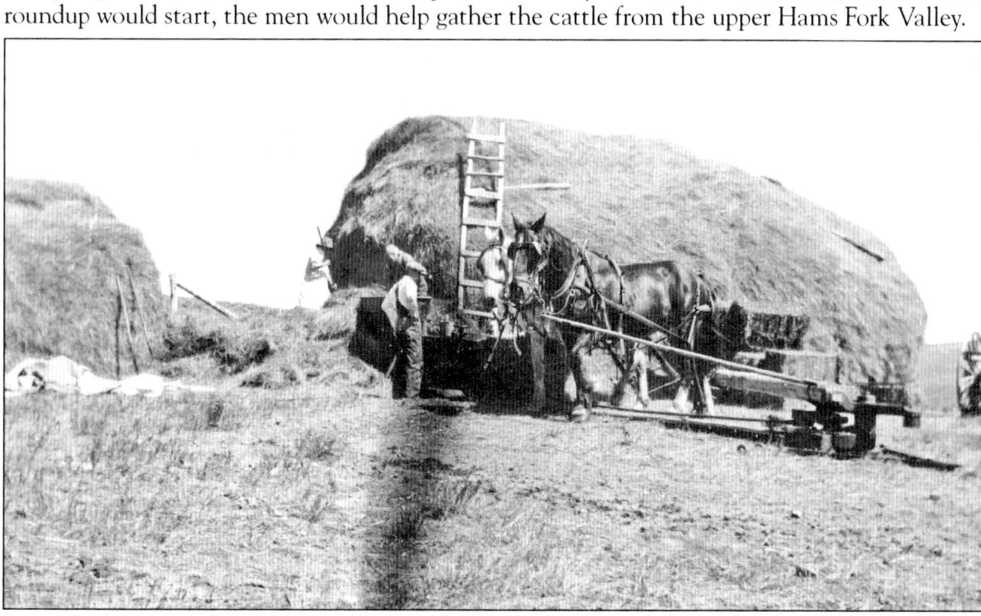

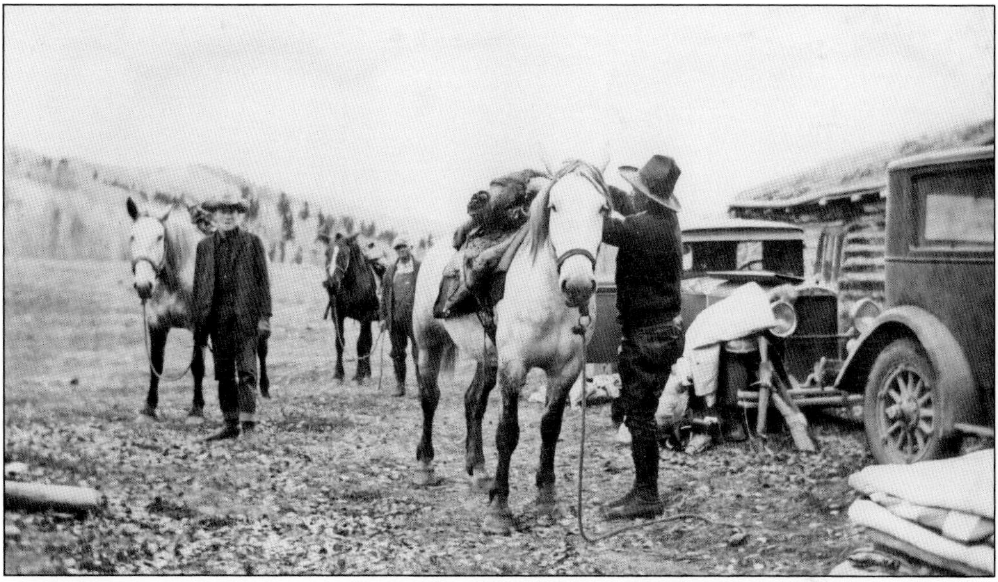

The favorite time of year for a lot of Kemmerer residents is fall. Fall is hunting season. In the early days, hunting had to be done from the back of a horse or on foot. The threat of snow usually stopped vehicles from traveling too far up the Hams Fork Valley. Pack horses or wagons were used to haul the needed supplies into the hunting camps. The pack horses were also used to haul out the big game animals that were killed. Hunting was a family event just like it is today. Everyone has his or her favorite hunting spot and keeps it a secret.

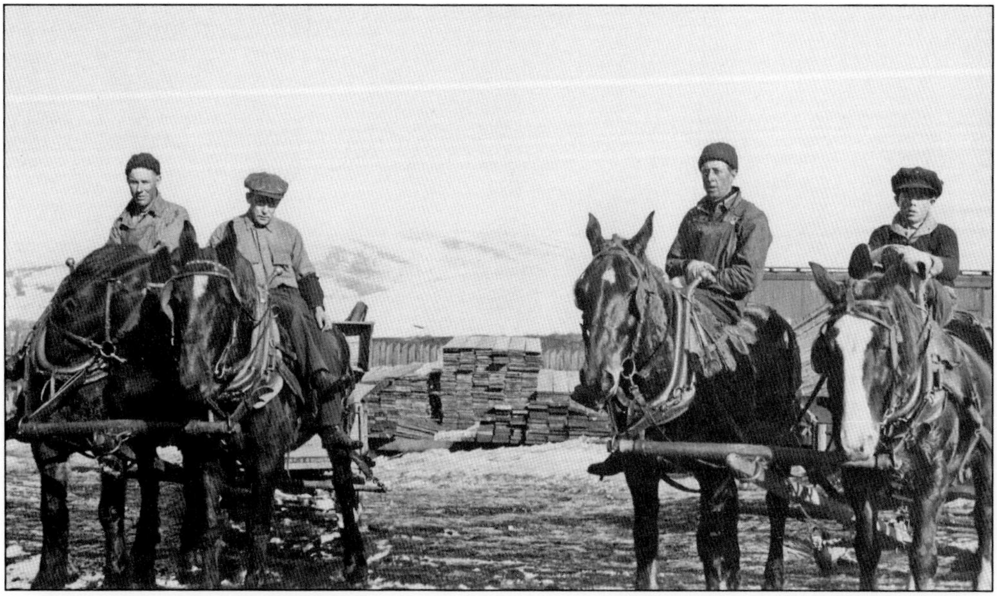

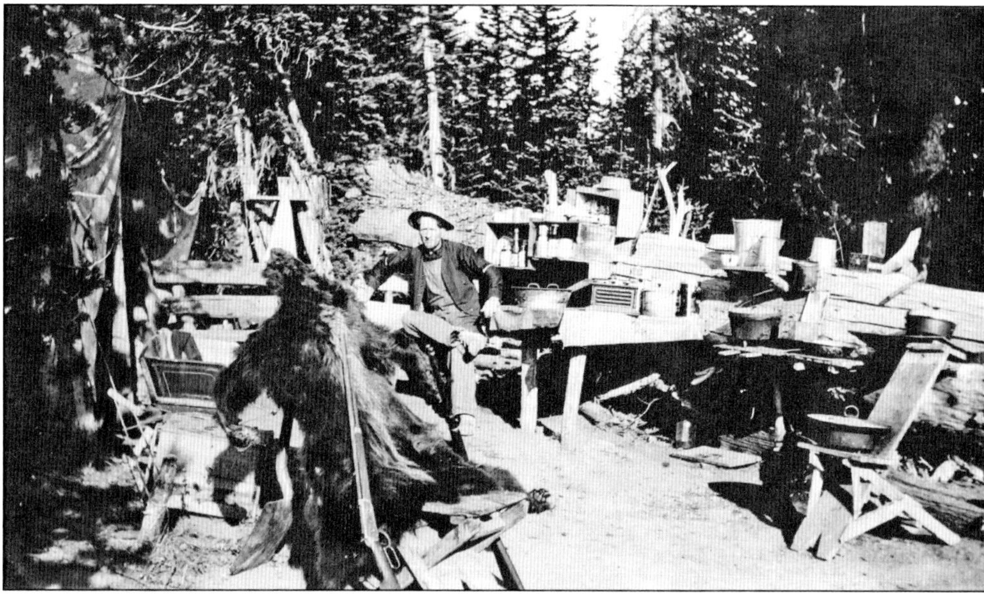

Hunting and fishing have always been favorite activities for the people in the area. Hunters have hunted to provide needed food for their families. In the early days, this was sometimes done by poaching. During hunting season, elaborate hunting camps were set up with handmade chairs and tables. Dutch ovens and a large supply of food, along with a warm, dry place to sleep, gave the hunters all the comforts of home. Fishing was a favorite summer activity. The fishermen would rise early to catch their limit of fish, eating that limit for breakfast. Resident fisherman paid $1.50 per year for licenses, and the limit was 20 fish or 15 pounds.

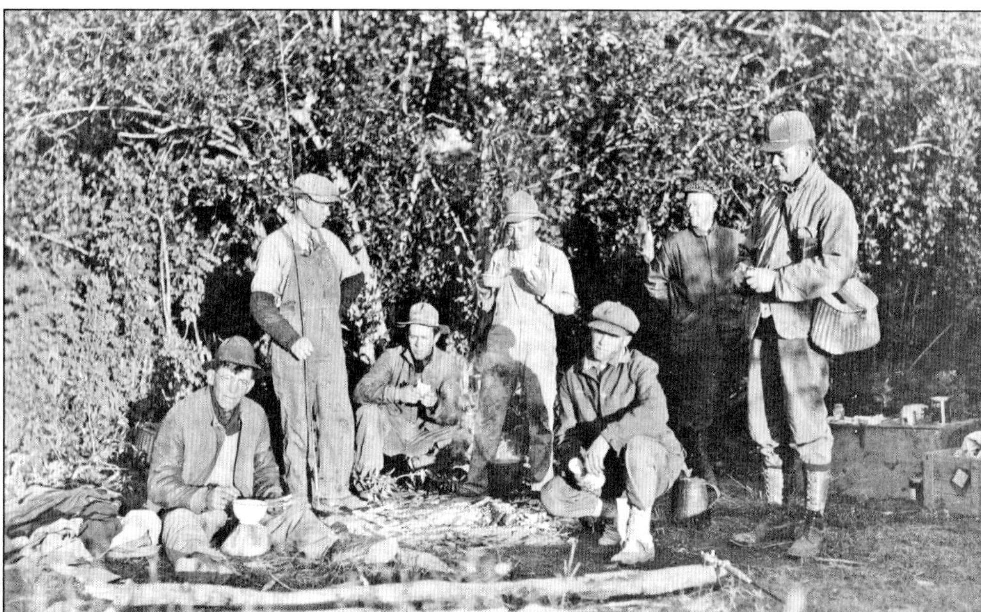

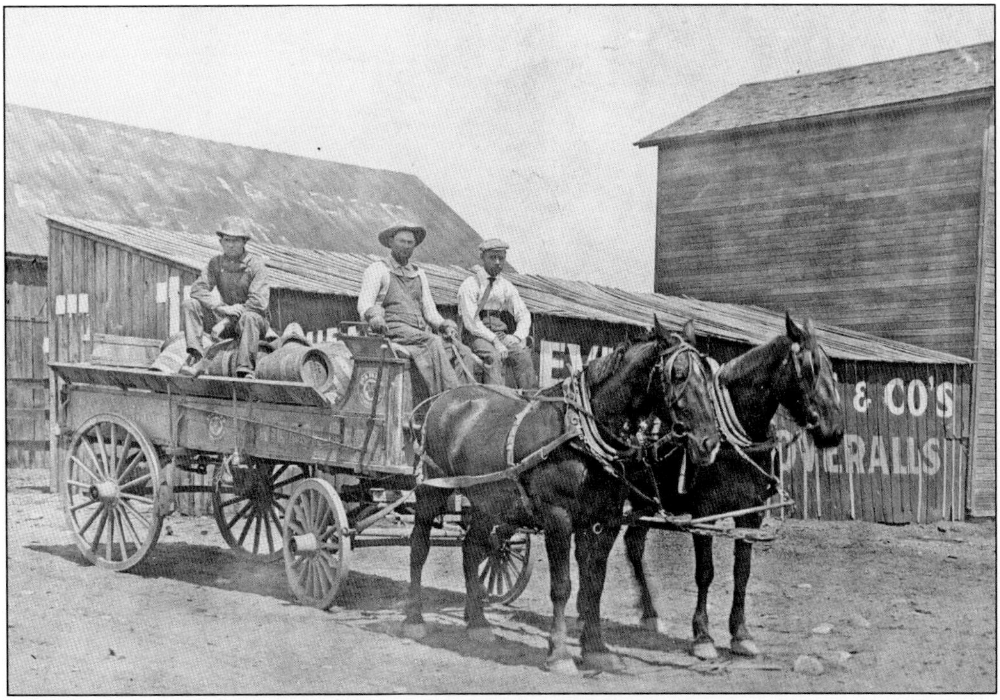

Before automobiles made their first appearance in Kemmerer, deliveries were made with wagons. The wagon in the photograph is a beer wagon that belonged to Lawrence Quealy, who owned the Frontier Saloon. The kegs of beer would be brought into Kemmerer on the railroad, and then Quealy's freight wagon would pick up the kegs and deliver them to the local saloons.

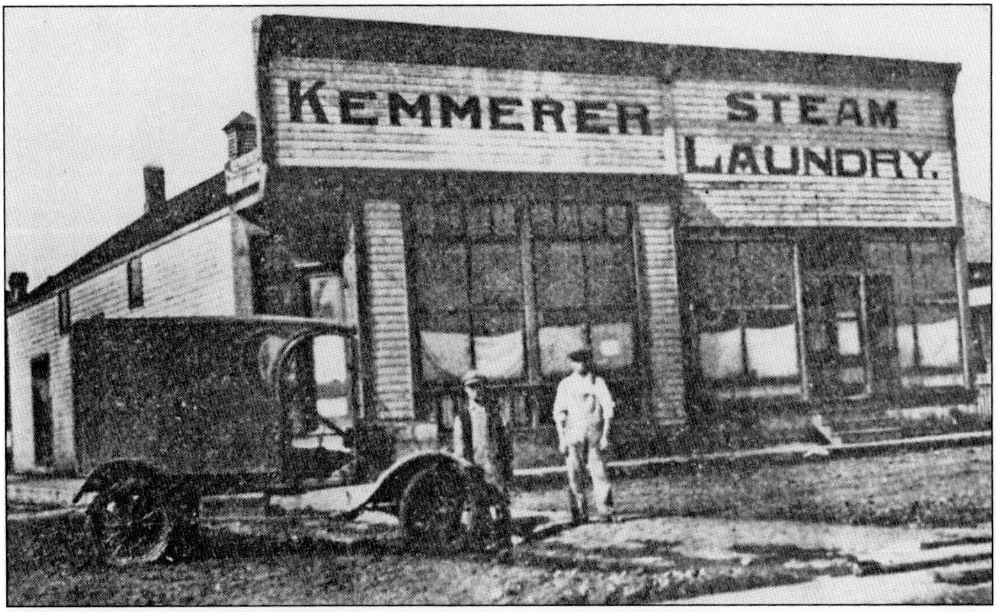

As Kemmerer grew, the need to give customers better service led to home deliveries. One of the first businesses in Kemmerer to have automobile delivery was the Kemmerer Steam Laundry. The laundry picked up and delivered to its local customers. This service made it easier for the customer and helped improve business.

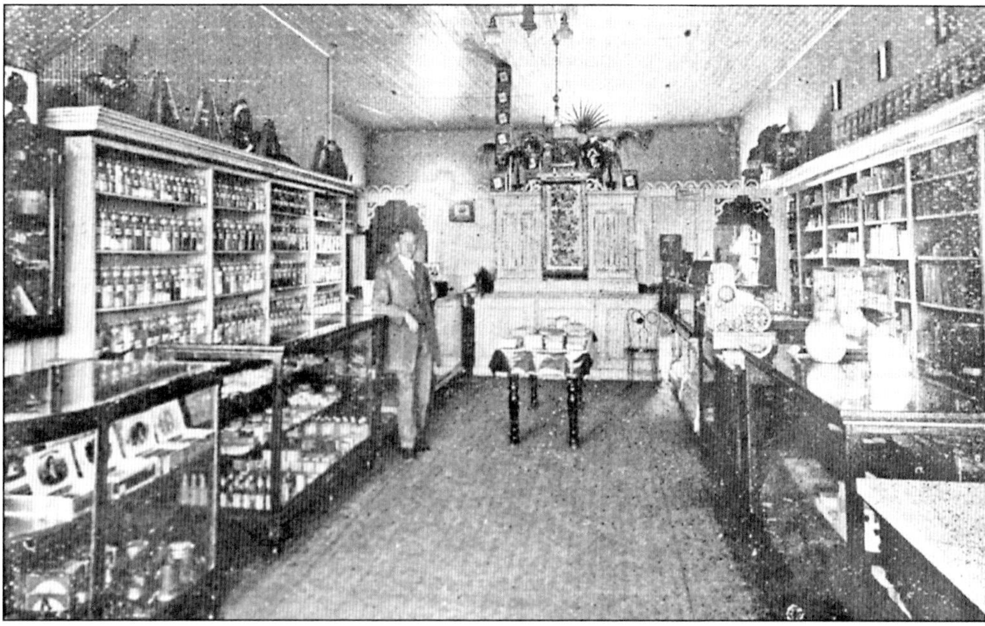

This photograph shows the interior of Kemmerer's first drugstore. Opened around 1905, the store was owned by Charles A. Christmas. The building was located on Pine Avenue. Christmas sold everything from medicines and home remedies to camera supplies and cigars. The store could provide most things a person would need.

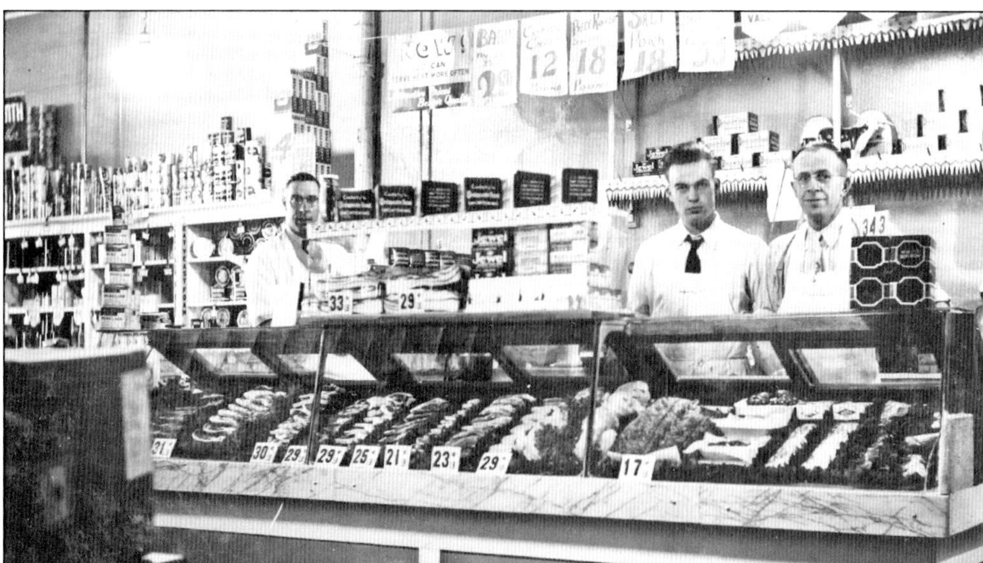

Kemmerer has had a number of different grocery stores. One of the early stores was the Safeway store, located on Pine Avenue where the current Ace Hardware store is. This 1940 photograph, from left to right, shows the store butchers Jim Buck, Franklin Barker, and Mr. Cameron.

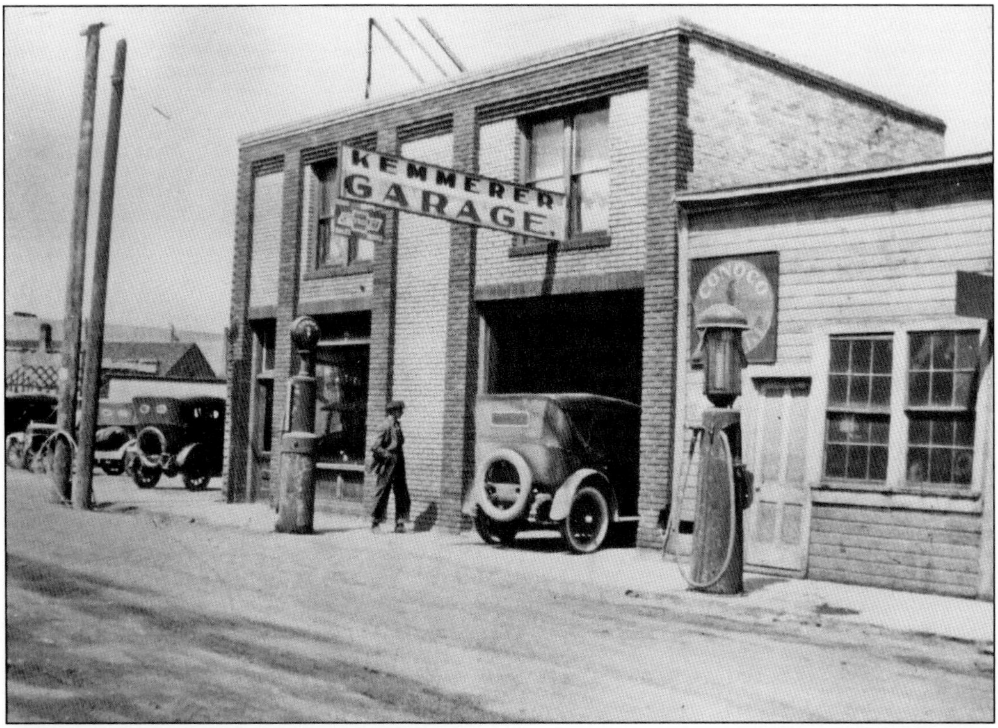

There were several garages in Kemmerer from 1915 to 1960. Originally the garages were blacksmith shops, and they gradually began to sell and repair automobiles. Some of the garages also sold gasoline. In 1923, Frank Hlachun owned the Kemmerer Garage located on Klondyke Street next to the present Eagles home. He advertised his garage as the home of quality cars with expert repairing. The Kemmerer Garage was also known for rigging the cars to carry the heavy loads of moonshine. In 1918, Platt Wilson opened the Lincoln County Garage next to the Union Laundry and across from the railroad depot. The building burned down in the early 1920s. Wilson rebuilt the garage at a different location and renamed it Platt Wilson Motor Company.

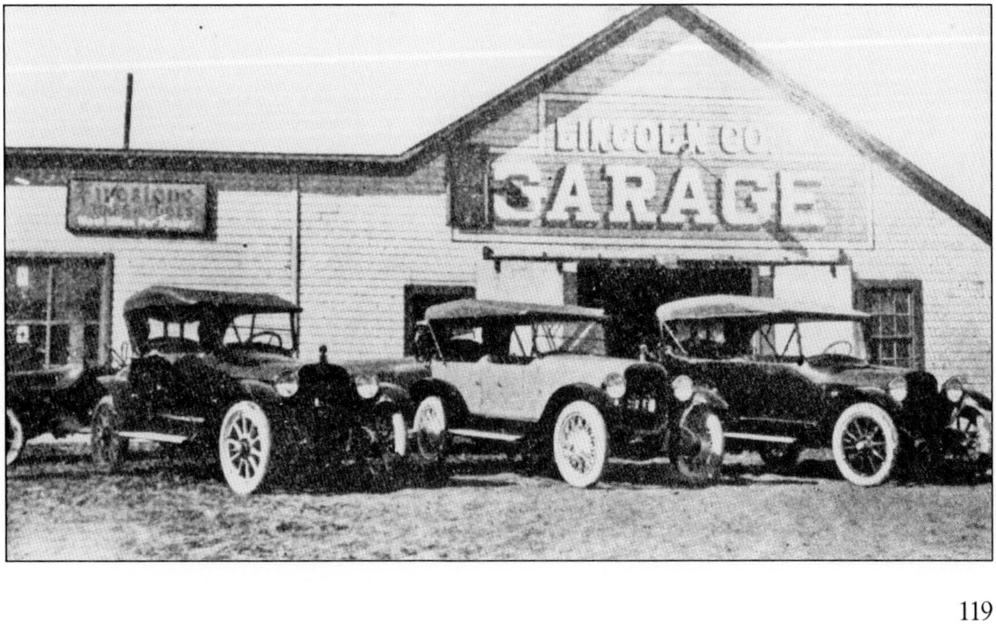

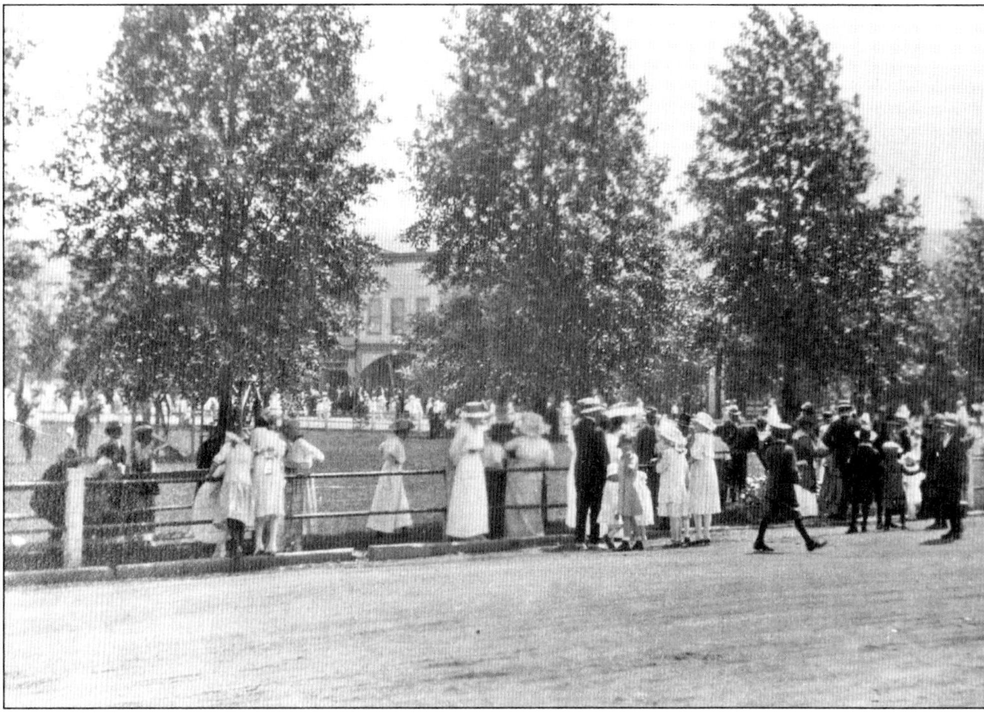

From its beginning in 1897 to the present, the Herschler Triangle Park has always been the center of social activity in Kemmerer. In the early 1900s, people would dress in their Sunday best and go to the park. All kinds of celebrations, from the Fourth of July to Labor Day, were held in the park. They would have picnics and play games during the celebrations. The local bands would provide music from the bandstand; the band would usually play into the early morning hours. Besides celebrations, other gatherings were held, like political speeches, the Frontier No. 1 mine explosion victims' funeral, and the return of the soldiers from World War I.

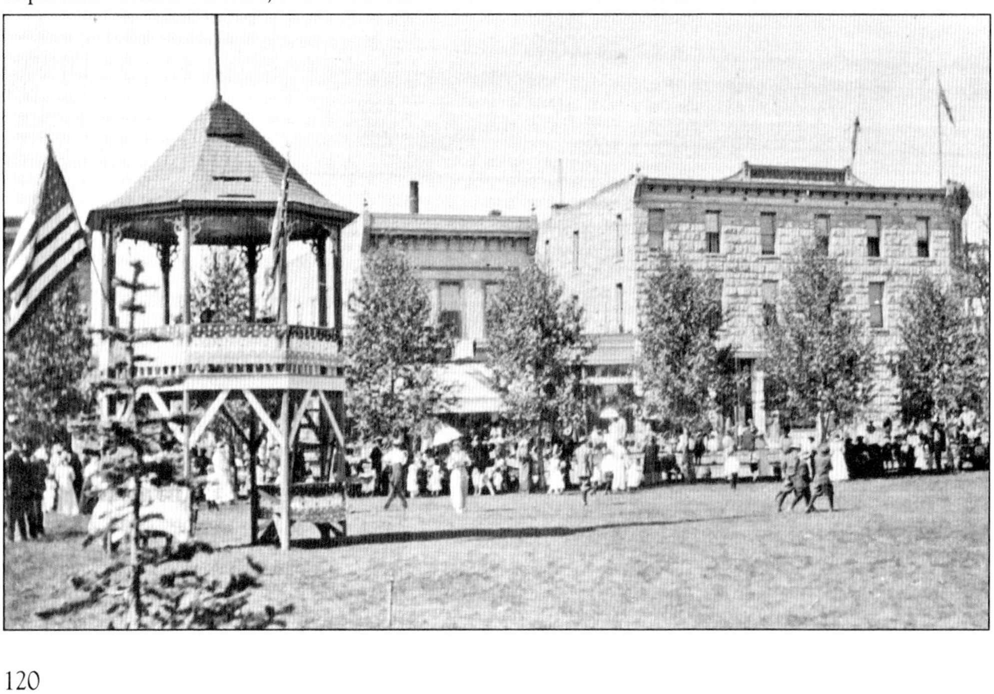

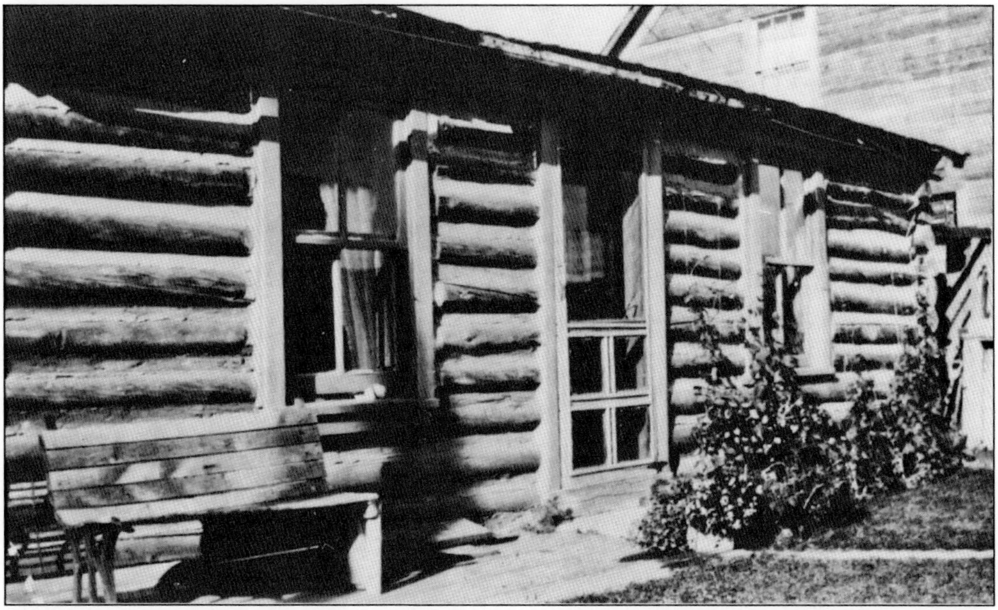

This photograph shows the oldest known home in Kemmerer, built in 1897 by Edwin Tyne Kirkwood. The logs for the home came from the upper Hams Fork Valley. Kirkwood brought the other material by freight wagon from Almy, Wyoming. The home had several owners over the years; it was torn down in 1959.

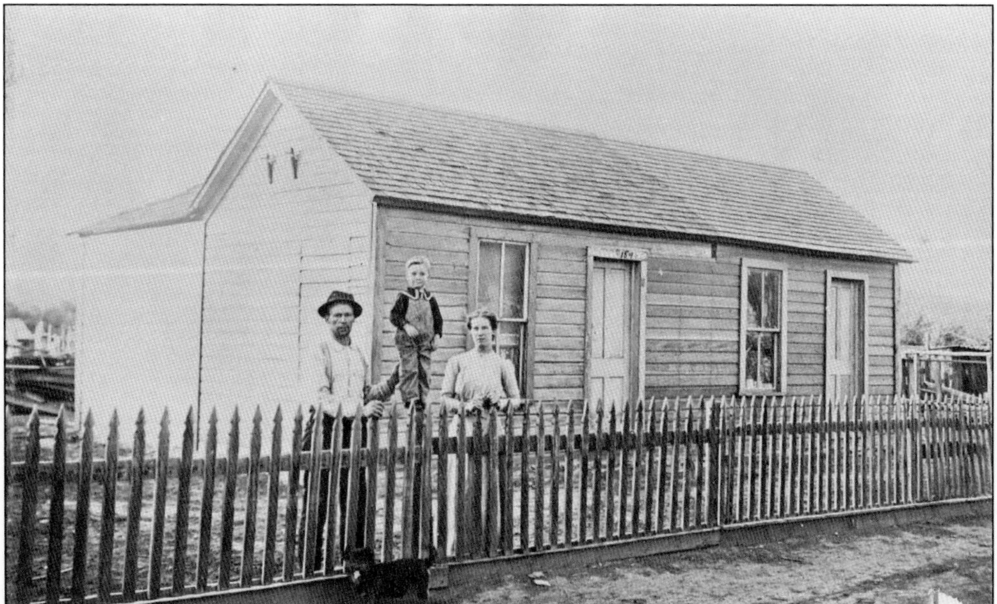

All of the homes located in Frontier were owned by the Kemmerer Coal Company. There were originally 16 homes; as the mine grew and more men were employed, the company built more homes. The homes had three or four rooms and a porch. Those on the north side of the river had no water, so they had to dig wells; later the company installed water pumps in the houses.

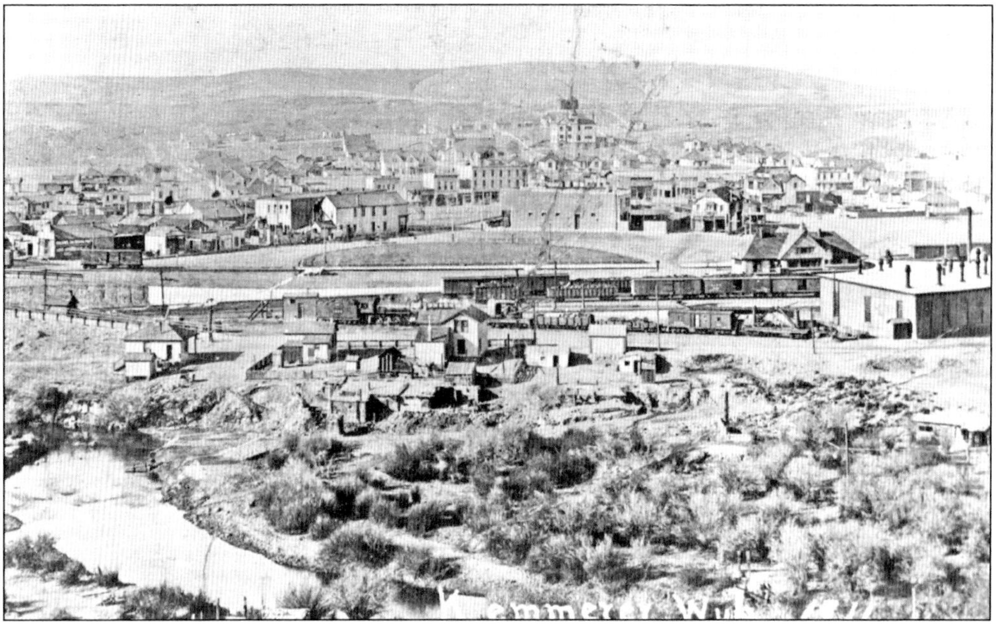

Over the past 111 years, Kemmerer has continued to grow and change but has remained a hardworking, close-knit community. It has become a center for large industry and home to the nation's deepest open-pit coal mine. Most of the original families that came to the Kemmerer area to work in the coal mines still have descendants living here. The employees who work in the coal mine, power plant, gas plants, area stores, city and government agencies, and school district and all the ranching families have helped contribute to the continued growth of the area and the success of Kemmerer. The photograph above shows Kemmerer c. 1925; below shows the town c. 1960.

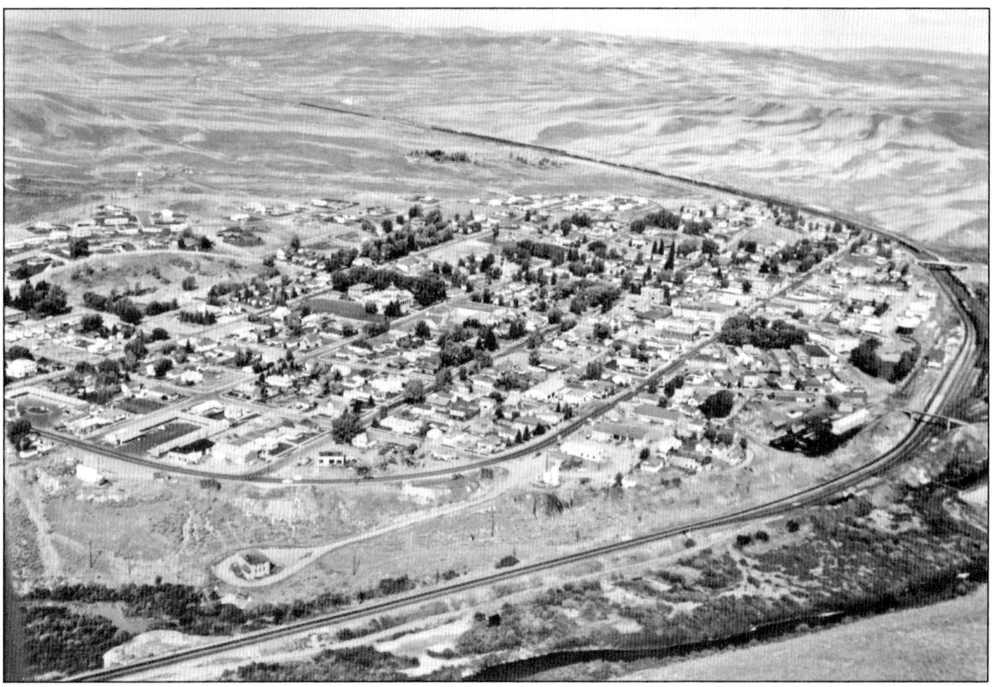

Seven

Fossil Country Museum

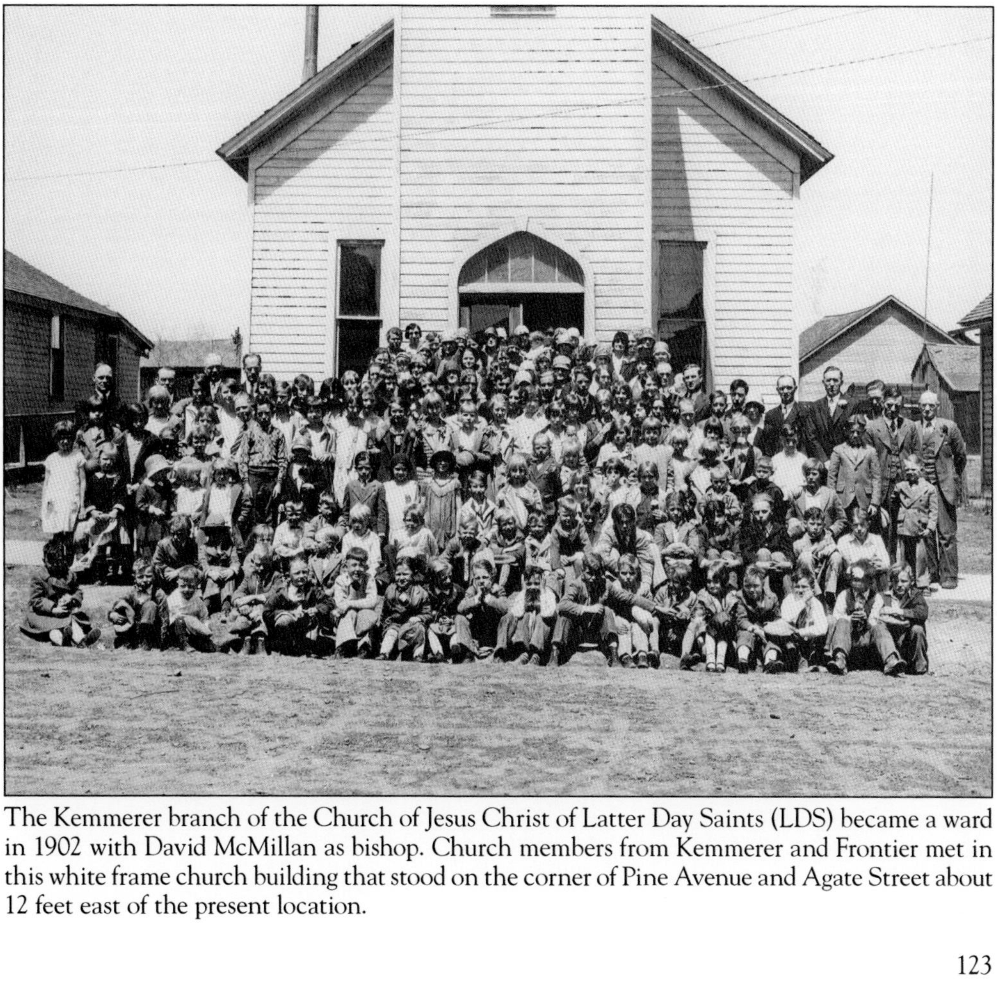

The Kemmerer branch of the Church of Jesus Christ of Latter Day Saints (LDS) became a ward in 1902 with David McMillan as bishop. Church members from Kemmerer and Frontier met in this white frame church building that stood on the corner of Pine Avenue and Agate Street about 12 feet east of the present location.

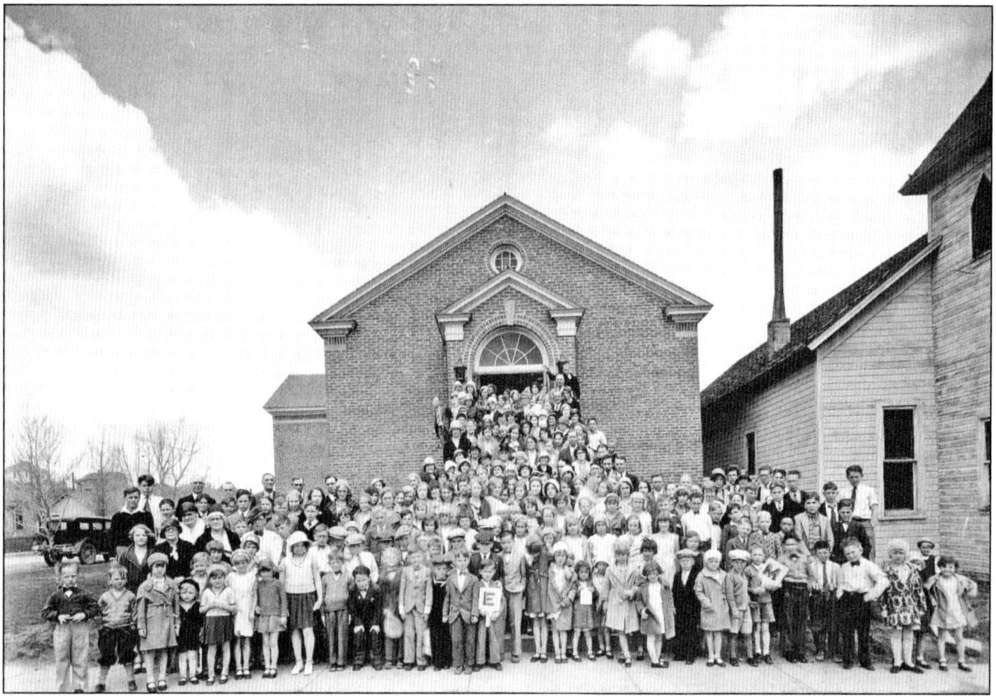

Church members attended the small white church from 1902 until 1930, when a new brick building was constructed (above). The new church was built next to the old white one. The construction on the new building was completed in late 1930; the new building was dedicated on January 4, 1931. It consisted of two floors, the lower floor containing a kitchen, heating plant, and classrooms. The main floor held the large chapel, an office, and a library. The chapel had seating for 250 people and choir seating for 50. After the brick building was dedicated, the wooden building was used as a recreation center for many church activities. The inside of the new church is pictured below.

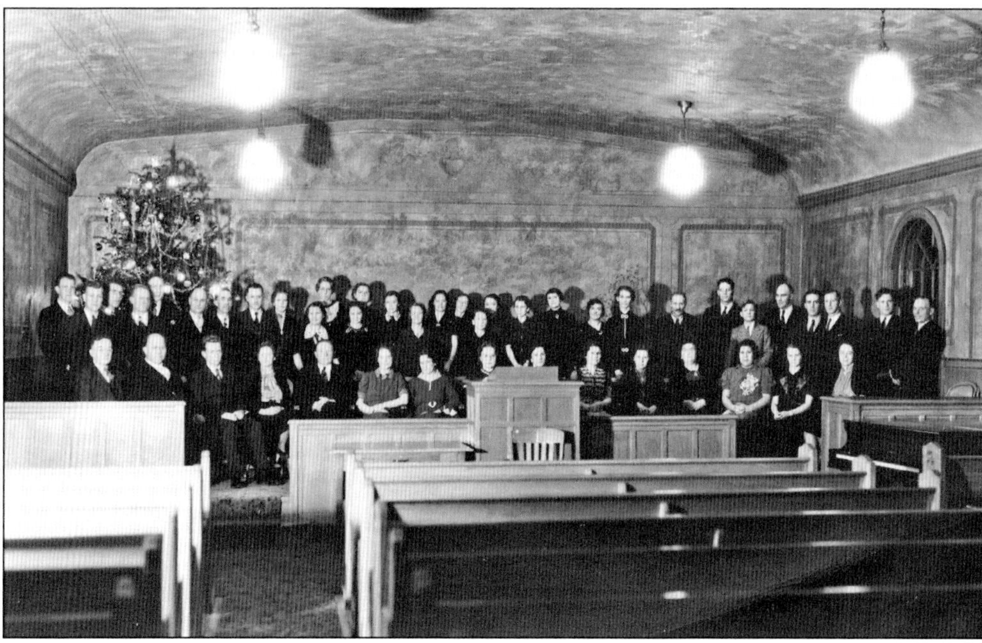

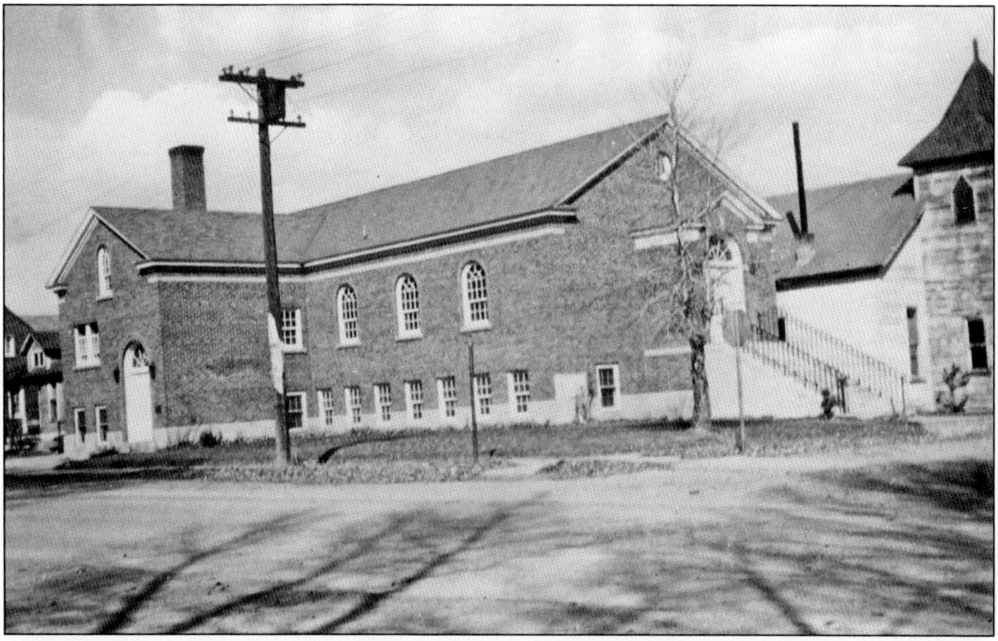

As the church congregation grew, there became a need for additional room. The old white church building was torn down, and in its place, a much needed addition was built. Ground-breaking on the new addition began on August 3, 1959. A large recreation hall, an updated kitchen, and additional classrooms were constructed. The chapel, which originally was faced north, was remodeled to face south. A large parking lot was also added in the back of the building. Church services were held in this building for 58 years. The last meeting was held in the building in 1988 before it was vacated for a larger one. The church is shown above in the 1950s and below in the 1960s.

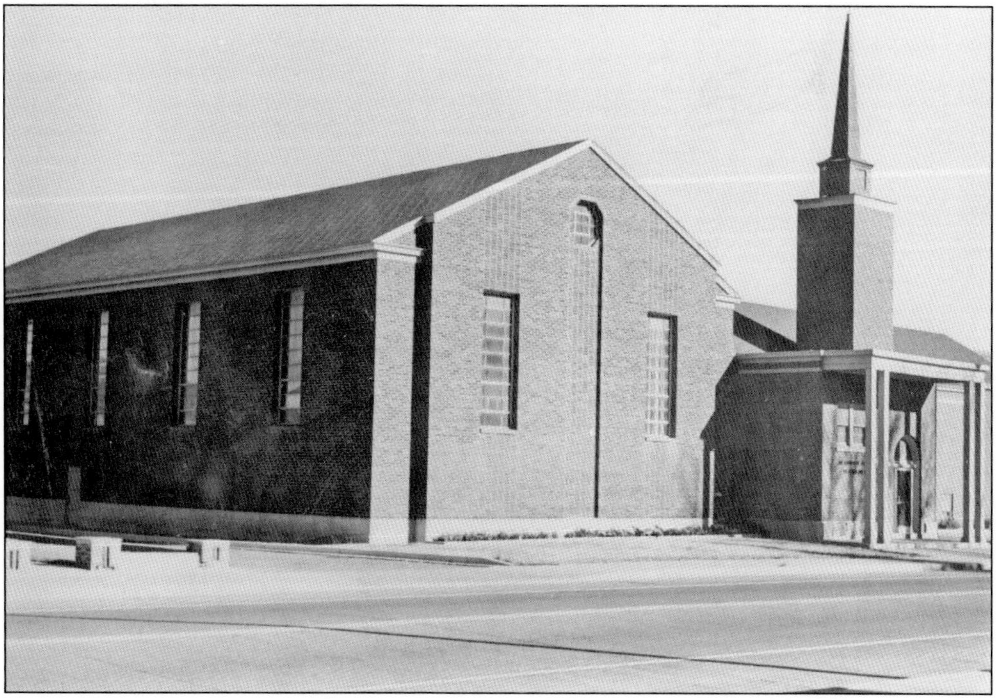

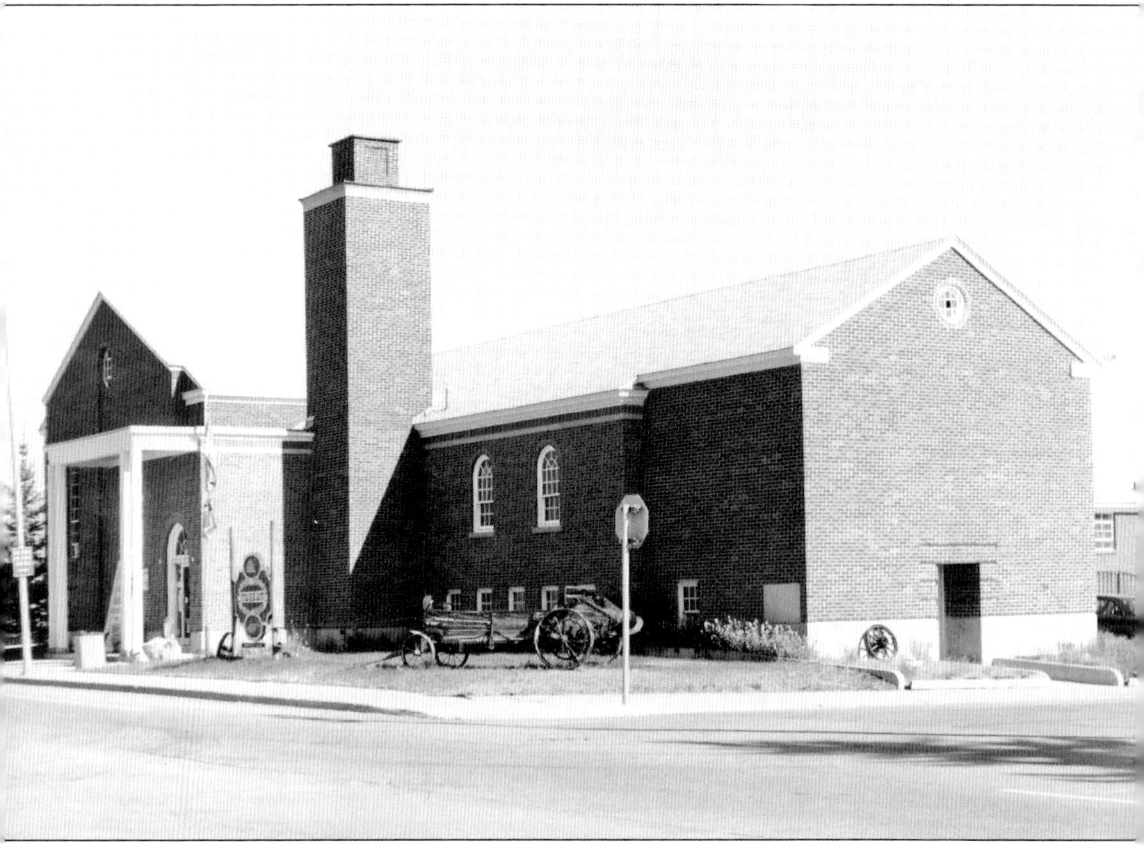

On January 13, 1988, Fossil Country Futures, Inc., a nonprofit tourism group, made a proposal to the LDS Church headquarters in Salt Lake City, Utah, for the building. The group proposed to use the empty building to create an education and resource center. The center would be referred to as the Fossil Country Museum. A $25,000 planning grant was awarded to the group for the writing of the museum plan. The proposal consisted of a donation from the LDS Church and a matching grant from the Wyoming Farm Loan Board. The parties agreed, and the purchase agreement was signed on July 29, 1988, by Nancy Peternal. Fossil Country Museum was dedicated on June 1, 1989, by state officials. The museum archives ,as well as the exhibits in the Heritage Gallery, have grown over the years. The most popular exhibits are the replica underground coal mine, bootlegging display, and the two-bodied lamb. The year 2009 will mark the 20th anniversary of Fossil Country Museum.

BIBLIOGRAPHY

Antilla, Alice. *First One Hundred Years Coal Mining—Hams Fork, Kemmerer, Lincoln County*. 2004.

Barrett, Dr. Glen. *Kemmerer, Wyoming: The Founding of an Independent Coal Town 1897–1902*. Kemmerer, WY: Quealy Services, Inc., 1975.

Essman, Ray. *Only Count the Sunny Hours: Madame Isabelle's Diary*. Winona, MN: Apollo Books, 1981.

Fagnant, Marcia. *Fossil Wyoming: A Trip through Time and Memories*.

Moonshine Madness. Kemmerer, WY: Fossil Country Museum, 1995.

Wright, Dorothy. *Kemmerer Wyoming, Digging for Western History*. Powell, WY: Pronghorn Press, 2007.

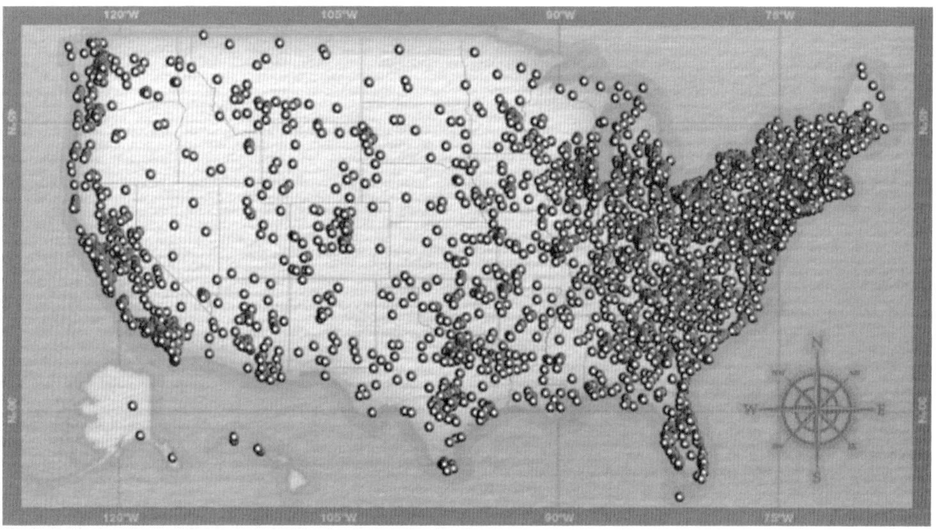